BONES

BONES

A STUDY OF
THE DEVELOPMENT AND STRUCTURE OF
THE VERTEBRATE SKELETON

BY

P. D. F. MURRAY, M.A., D.Sc.

With an Introduction by
B. K. HALL
Department of Biology, Dalhousie University,
Halifax, Nova Scotia, Canada

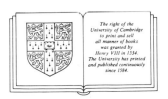

The right of the
University of Cambridge
to print and sell
all manner of books
was granted by
Henry VIII in 1534.
The University has printed
and published continuously
since 1584.

CAMBRIDGE UNIVERSITY PRESS
Cambridge
London New York New Rochelle
Melbourne Sydney

Published by the Press Syndicate of the University of Cambridge
The Pitt Building, Trumpington Street, Cambridge CB2 1RP
32 East 57th Street, New York, NY 10022, USA
10 Stamford Road, Oakleigh, Melbourne 3166, Australia

First published 1936
Revised as a paperback with an introduction 1985

Printed in Great Britain at the University Press, Cambridge

Library of Congress catalogue card number: 84–29392

British Library cataloguing in publication data
Murray, P. D. F.
Bones: a study of the development and
structure of the vertebrate skeleton. –
(Cambridge science classics)
1. Skeleton
I. Title
596′.01852   QP301
ISBN 0 521 31549 2

UP

EQUALLY
TO MY WIFE AND
τῷ Φερεοίκῳ

CONTENTS

PREFACE

This book does not attempt to be comprehensive in its survey of the facts and theory of skeletal structure and development; that would require a book many times the size of this. I have merely tried to discuss in a brief space certain aspects of the development and structure of the skeleton, and in particular the question of the relation between the structures seen and the morphogenic factors intrinsic in the elements on the one hand, and those, mainly of a mechanical nature, which act upon it from without, on the other. In doing this, I have tried to present a fair picture of the present position of a controversial subject and to indicate, sometimes perhaps by daring to venture a little beyond the safe entrenchments of established fact, what seem the probable directions of future theoretical development and the lines of most profitable investigation. The main mass of the book is in the first four chapters, which deal with the embryonic and post-embryonic development and with the structure of the normal and modified bony skeleton considered in relation to the function it fulfils. In a brief fifth chapter cartilage is dealt with from the same point of view. The sixth is devoted to a consideration of the mechanisms by which mechanical forces find expression in the architecture of bone, and the seventh is a very brief summary.

Thanks are due to Dr H. B. Fell, Dr A. Glücksmann, Dr W. Jacobson, for a great deal of helpful advice and criticism, and for permission to mention their unpublished works, but particularly to Mr C. Mason, of the Cambridge Instrument Company, who devoted a great deal of time and trouble to assisting me with the mechanical passages, and to my wife, works, but particularly to Mr C. Mason, of the Cambridge Instrument Company, who devoted a great deal of time and trouble to assisting me with the mechanical passages, and to my wife, who helped very materially in a secretarial capacity. But of course none of these is responsible for any of the faults which the book may contain.

P. D. F. M.

September, 1936

*Research in the development and structure of the skeleton
since the publication of Murray's* Bones

Introduction

It is half a century since Cambridge University Press published
*Bones: A Study of the Development and Structure of the Vertebrate
Skeleton* by P. D. F. Murray. At the time of its publication
Murray, just 36 years old, was Smithson Research Fellow at
the Strangeways Research Laboratories in Cambridge,
England. Scion of a family whose accomplishments spanned
scholarship, administration and politics (his uncle was Sir
Gilbert Murray, the renowned classical scholar and Regius
Professor of Greek at Oxford; his father, Sir Hubert, was for
32 years Lieutenant-Governor of Papua, and his grandfather,
Sir Terence, was President of the Legislative Council of New
South Wales, Australia), Murray won a University Medal
upon graduation from Sydney University in 1922, proceeding
to take a BSc at Oxford in 1924 and a DSc at Sydney in 1926,
the latter based on the first of a life-long series of studies on the
development of the skeleton. He pioneered the technique of
grafting embryonic skeletal tissues onto the chorioallantoic
membranes of host embryos as reported in three papers
published in 1925 with J. S. (later Sir Julian) Huxley, papers
published when Murray was just 25 years old. His interest in
the relative roles of extrinsic and intrinsic factors in the ac-
quisition of skeletal form developed early, as evidenced by an
extensive review published with D. Selby in 1930. This topic
formed a major theme of *Bones*. As Murray indicated in the
Preface: 'I have tried to discuss...the question of the relation
between the structures seen and the morphogenic factors
intrinsic in the elements on the one hand, and those, mainly of
a mechanical nature, which act upon it from without, on the

other.' He was to review the topic again in the 1947 issue of the *Annual Review of Physiology*. Murray went on to publish on the origins of haematopoietic centres and fibrillation of the heart in the early embryo, on vitamin C deficiency in fracture repair and wound healing (his major work of the Second World War), and on the mechanical evocation of secondary cartilage from periosteal cells on avian membrane bones. *En passant*, he published one of the first integrative undergraduate textbooks on biology, a masterly fusion of zoology, botany, anatomy and physiology which was very widely used by both biology and medical students. A complete bibliography of Murray's papers may be found in the obituary published by the Australian Academy of Science (Rogers, 1968).

The publication of *Bones* marked a milestone in skeletal biology. No longer could the structure of the skeleton be viewed solely as a response to mechanical factors impinging upon it. Murray showed, in just 181 pages of succinct analysis and flowing prose, and using the experimental work which he and others had embarked upon, how the skeleton was shaped by a delicate balance of intrinsic and extrinsic factors. All three major 1937 reviews of the book (*Nature* **139**, 1036–7; *Quart. Rev. Biol.* **12**, 119; *J. Anat.* **71**, 582) emphasised this new paradigm. So convincing was Murray's case for intrinsic control of basic form that the reviewer in *Nature* commented that one 'cannot help being impressed by the inherent formative properties of the cells themselves. It is difficult to refrain from endowing them with potentialities that make further scientific investigation impossible' (p. 1037). Murray's insights into these 'inherent formative properties' led him to exactly the opposite conclusion. He ends the book with the positive affirmation of the experimental scientist that 'the real need is...for investigations of the mechanism which underlies the adaptations known to occur', and that such problems 'are probably not inaccessible to the ingenious mind and skilful technique of the biological experimenter' (p. 181). Indeed, they were not, as shown by Murray's subsequent work and by the explosion of experimentation since the publication of *Bones*.

Murray's book continues to be cited as a benchmark in the field. Its influence extends well beyond biology; it is cited in the primary literature of anthropology, anatomy, dentistry, engineering, medicine and veterinary science. In this brief introductory essay I want to present an overview of the progress which has been made in understanding what Murray called 'the obscure processes of skeletogenesis'. All that has happened since 1936 cannot possibly be covered. Rather, I will evaluate the major topics discussed by Murray and briefly summarise where we now stand.

The origin of skeletal cells

Murray began the book with a discussion of the development of the cartilage which forms the primary model for the long bones of the limbs, bones which form by endochondral ossification (mammals) and/or by sub-periosteal ossification (birds). He only made passing reference to the craniofacial skeleton and to the vertebrae (pp. 2, 27, 29, 175), for virtually nothing had been done on these regions of the skeleton. We now have data on all regions of the skeleton for a large number of vertebrates. Much of the data for many groups is still only descriptive, especially for fishes, reptiles and many mammals, but much other deals with underlying mechanisms. The Appendix gives an indication of the range of general works available. It is not exhaustive but was simply chosen by scanning the bookshelves of my office. For a further topically organised list of references see Ham and Cormack (1979).

Murray opens with a discussion of condensation of

mesoderm – the first evidence of a future skeletal element. A condensation consists of preskeletal cells and occurs at the site of future skeletogenesis. Much is now known about the condensation process, both *in vivo* and *in vitro*. Three mechanisms are operative: division of cells *in situ*, aggregation or migration of cells towards the future skeletal site and suppression of cell death at the site (Hall, 1978, chap. 5; Bowen and Lockshin, 1981; Ede, 1983; Fyfe and Hall, 1983; Solursh, 1983). These local accumulations of cells form a skeleton themselves – what Grüneberg (1963) has called the membranous skeleton of the embryo (see also Ede, 1983).

Condensation places the mesenchymal cells sufficiently close together that they can begin to interact with one another. Such interactions are confined to cells at particular stages of differentiation (raising the possibility that they might be consequences not causes of such differentiation), and often involve extracellular matrix products such as collagen, or alterations in intracellular molecules such as cyclic AMP. Rounded cell shape is critical for such differentiative interactions (Abbott and Holtzer, 1966; Glowacki, Trepman and Folkman, 1983; Solursh, 1983), but we still have a long way to go in understanding how altered cell shape relates to the initiation of skeletogenesis. Evidence from analysis of mutants in which skeletal development is affected indicates that attaining a critical condensation size is often, but not always, necessary for the initiation of chondrogenesis or osteogenesis (Grüneberg, 1963; Grüneberg and Wickramaratne, 1974).

A major aspect of mesenchymal condensation which has come to light since Murray's book is the observation that skeletogenic mesenchymal cells migrate to their final site from another region of the embryo. Although these cells condense *in situ*, they do not arise *in situ*. Future vertebral cells (the sclerotome) move from the somites to take up positions adjacent to the spinal cord (Hall, 1977). Cells of the craniofacial skeleton arise in the neural crest, a ridge of cells along the dorsal surface of the neural tube; i.e. some of the skeleton arises not from *mesoderm* but from *ectoderm* – heresy in the 1930s but common parlance today. These neural crest cells leave the neural tube to undergo extensive migrations, first to create the craniofacial processes themselves and then to differentiate into the cartilage, bone and dentine of the chondro- and osteo-cranium, visceral arch skeleton and teeth (Noden, 1980, 1983; Le Douarin, 1982). Future craniofacial myogenic cells arise *in situ*. In the developing limb bud it is the future skeletal cells which arise *in situ* and the myogenic cells which arrive from outside, in this instance from the somatic mesoderm (Cheval-lier, Kieny and Mauger, 1977; Christ, Jacob and Jacob, 1977). The mesenchyme of the regions in which the primary skeleton and its musculature develops is therefore heterogeneous, although much of the experimental work of the 1960s and 1970s especially that on chondrogenesis and myogenesis *in vitro*, treated the mesenchyme as a homogeneous population of equivalent cells. Murray wrestled with this problem on the first page of his book. Are the cells in the condensations pre-determined or are they indifferent with respect to their differentiative fate? Both their distinctive embryological origins and the interactions which occur between them argue for determination having occurred by the time that conden-sation takes place.

The shape of that initial condensation is not randomly generated but is established with remarkable precision well before any cartilaginous or bony matrix has been deposited, as elegantly demonstrated by Noden (1983) and by Hinchliffe (1977). (See Hinchliffe and Johnson (1980) for a thorough discussion.) Within that condensation the shape of the future skeletal rudiment is foreshadowed (discussed on pp. 6–10 of

Bones). Is that shape generated by positional effects as postu-
lated by Wolpert (1978), by mechanical contractile forces
associated with cell movement (Oster, Murray and Harris,
1983), by morphogens (Turing, 1952), or by differential
growth (Wolpert and Hornbruch, 1981)? Although these
mechanisms vary greatly, they have as a common denom-
inator the concept of intrinsic control, independent of adjacent
tissues and of the biomechanical factors which mould so much
of the subsequent shape of the skeleton (see pp. *xxiv–xxxii*).

Thus, the skeleton arises from cells which often have to
migrate to the site of skeletogenesis, there to form condensations, a process which allows similarly determined cells to
interact with one another. Such future skeletogenic cells may
be distinguished from cells outside the membranous skeleton
by this determination for chondrogenesis or osteogenesis. As
Murray points out (pp. 3–5), cartilage- and bone-forming cells
are not separate cell lineages. He cites the classic work of Fell
(1933) on formation of cartilage from endosteal cells *in vitro*.
Murray himself went on to document the formation of
cartilage from otherwise osteogenic periosteal cells of membrane bones (Murray, 1957, 1963; Murray and Smiles, 1965).
Such secondary cartilage is now routinely recognised in histo-
logy texts as a major class of cartilage and has a book devoted
to it (Beresford, 1981). These periosteal progenitor (stem) cells
are bipotential, forming cartilage when subjected to mechanical
forces, but otherwise forming bone. The existence of secondary
cartilage is but one example from the many studies which have
firmly established the concept of the skeletogenic or osteo-
chondrogenic stem cell, and which have shown that cartilage
and bone (and often connective tissue as well) are an inter-
related family of tissues which arise by modulation from such
stem cells (Bassett and Ruedi, 1966; Hall, 1970, 1978; Owen,
1970; Beresford, 1981).

Cells *outside* the condensations may also have skeletogenic potential. Murray (pp. 3–4, 151–2) discusses the classic study of Huggins (1931) which demonstrated that urinary bladder epithelium induces bone from connective tissue cells in some tissues but not in others. Urist, Dowell and Hay (1968) have confirmed this spatial heterogeneity of responsive cells and the pioneering work of Friedenstein in the USSR has identified two classes of osteoprogenitor cells in the adult, but outside the skeleton, one which is already determined for osteogenesis (his DOPC, determined osteogenic precursor cells) and one which has to be induced before it will deposit bone (his IOPC, inducible osteogenic precursor cells: Friedenstein, 1973, 1976; Owen, 1978). As with cells within condensations, such precursors can also form fibrous tissue, i.e. they are bipotential, a phenomenon especially well illustrated by intratendinous ossification (see Murray, pp. 114–17 and 153–4; Wong and Buck, 1972; McClure, 1983). Proliferating epithelia are potent inducers of the subset of adult cells capable of responding to them (Wlodarski, 1969; Anderson, 1976; Hall 1983 a). Epithelia also play a critical role in evoking chondro- and osteogenesis *within* condensations (Hall, 1983 a, b), highlighting the commonality of epigenetic interactions in embryonic and post-foetal skeletogenesis. Other potent inducers of ectopic skeletogenesis have been discovered since 1936, notably bone morphogenetic protein, a glycoprotein of molecular weight 17 500 resident within bone matrix (Urist, 1980, 1983). This may or may not be the same factor as that isolated by Reddi (1983), a protégé of Huggins who started the field 55 years ago. There are two common denominators which emerge when considering intra- and extraskeletal osteogenesis and chondrogenesis: the existence of competent cells and the presence of a triggering mechanism to activate those cells.

Intra- and extraskeletal regulation of osteogenesis also act together in the osteoclast. Murray had very little to say about these bone-resorbing cells (pp. 162–6). Without doubt it is our changing view of the osteoclast which has most revolutionised the study of the skeleton. It was dogma for a century that osteoclasts and osteoblasts arose from progenitor cells *within* the skeleton, indeed that they were modulations of the *same* progenitor cells. The treatments underlying hormonal control of bone growth, remodelling and decay, and the treatment of metabolic bone diseases, started with this premise. Over the past ten years our conception of osteoclast origins and activation has changed dramatically, for it is now known that the osteoclast arises *outside* the skeleton, from precursors circulating within the blood stream, that these cells have to migrate to the skeleton and that their resorptive action, long known to be correlated with cycles of deposition of bone, is so coupled because osteoblasts release factors which activate quiescent osteoclasts or their precursors (Owen, 1970, 1978; Hall, 1975; Ham and Cormack, 1979; Rodan and Martin, 1981; Mundy *et al.*, 1982; Malone *et al.*, 1982; Parfitt, 1984). Osteoblasts and osteoclasts are separate cell lines, united not by commonality of origin, but by functional interaction. This interaction is reflected in the current *in vitro* and grafting approach to osteoclast function, which is to co-culture or graft osteoclasts with bone as a matrix to be resorbed and with osteoblasts to provide the diffusible activating factors (Krukowski and Kahn, 1982; Chambers, 1982; Holtrop, Cox and Glowacki, 1982; Zambonin Zallone, Tetti and Primavera, 1984). Indeed, heterogeneity of osteoclast populations is beginning to emerge, as in the recent studies of Liu and Baylink (1984) and of Marks (1983). Is such heterogeneity based on separate races of osteoclasts or on specificity of activation? The documentation of such heterogeneity – fibroblasts with skeletogenic potential versus those without, site-specific populations or activation of osteoclasts, or the non-equivalence between a chondrocyte which goes to make up a wrist element and one which makes up a digit or humerus – is a major future task in our continuing search for the origins of skeletal cells and the control of their differentiation.

Morphogenesis of cartilage

A further major contribution made by Murray in *Bones* was the marshalling of evidence and arguments to show that the basic form of long bones – the relation of length to width, cross-sectional shape, basic longitudinal curvature, shape of major condyles – was not an adaptive response to mechanical influences, but rather was intrinsically programmed within the cells of the cartilaginous primordium (pp. 3–42).

From the mid 1920s on, cartilaginous rudiments were being organ cultured or grafted in isolation from muscular, nervous and/or vascular influences, and in isolation from influences of the cartilaginous rudiment with which they normally articulated. Such cultured and grafted cartilages formed skeletal elements of surprisingly normal morphology. Murray especially emphasised the normality of the shape of the shaft (pp. 4–6), articular structures (pp. 13–17 and figs. 2 and 3) and the major grooves and tuberosities which would, in real life, have accommodated ligaments and muscles (pp. 18–19). Dame Honor Fell, long-time Director of the Strangeways Research Laboratories where much of this early work was done, has reviewed the organ culture experiments (Fell, 1969, 1976); Felts (1961) has reviewed the grafting experiments; and recently Thorogood (1983) has provided a comprehensive overview of cartilage morphogenesis, including an evaluation of Murray's contributions, and discussions of the relationship between intrinsic and extrinsic factors in cartilage morphogenesis and between morphogenesis and growth. Thorogood identifies seven factors which impinge upon cartilage morphogenesis, viz., orientation of cells within the condensation, cell division, cell hyperplasia, secretion of extracellular matrix, growth of the cartilage, physical constraint of adjacent tissues and external mechanical stimuli. The fundamental shape of the cartilage is influenced by all of these factors but the basic intrinsic control lies within the condensation and particularly in the orientation of the cells within the condensation (Ede, 1971; Ede, Hinchliffe and Balls, 1977; Trelstad, 1977; Ede and Wilby, 1981; Thorogood, 1983). Differential adhesion of cells within the condensation plays a major role in that orientation (Ede, 1983). Somewhat later, the constraining influence of the perichondrium also comes into play (Wolpert, 1981).

That the behaviour of prechondrogenic cells within condensations is both intrinsic and cartilage-type specific is dramatically illustrated by the experiments of Weiss and Moscona (1958) where limb or scleral chondrogenic mesenchyme, upon dissociation into single cells and subsequent cell culture, forms rods or sheets of cartilage, respectively – forms which are equivalent to the shafts of long-bone primordia or to sheets of scleral cartilage. Such specific reassociation requires cell–cell interaction, for diluting the cell cultures with fibroblasts destroys the morphogenetic specificity of the reaggregate (Archer, Rooney and Wolpert, 1983). It is the surfaces and superficial extracellular matrices of the cells in the condensation that we will need to study if we are to understand the morphogenetic individuality of determined skeletogenic cells. Intrinsic control of skeletal growth can also be traced back to the condensation (Hall, 1982; Hinchliffe and Johnson, 1983), as is especially well shown when anlage of skeletal elements are grafted into hosts whose skeletons have different growth rates, final sizes or shapes (Twitty and Schwind, 1931; Iten and Murphy, 1980). More such studies are required.

Chondrones

Murray described chondrones as a basic structural unit within cartilage (pp. 136–48); each chondrone was thought to consist of a group of several chondrocytes encircled by a common fibrous network, with groups of chondrocytes similarly united by fibres until the entire cartilage came to constitute a linked set of chondrones (fig. 41, p. 140). The existence of such structural units has not stood the test of modern-day transmission and scanning electron microscopy (Stockwell, 1979; Boyde and Jones, 1983), so that chondrones are rarely mentioned these days. When they are, it is in relation to degenerative changes of cartilage in ageing or in pathology (Sokoloff, 1969; Stockwell, 1979). Unlike the osteon, which is the structural unit of bone, the chondrone has not been verified as a basic structural unit, despite our very extensive knowledge of the structural basis of cartilaginous extracellular matrix – cartilage-specific collagen and proteoglycan, glycosaminoglycan structure and attachment to collagen, and aggregation of matrix products (Lash and Vasan, 1983; Mayne and von der Mark, 1983).

Although chondrones have not been confirmed as the structural unit of cartilaginous extracellular matrix, considerable heterogeneity of the matrix does exist, especially when the territorial (pericellular) extracellular matrix around each chondrocyte is compared with the extraterritorial matrix between chondrocytes (Eisenstein, Sorgente and Kuettner, 1971, 1973; Kashiwa, Luchtel and Park, 1975; Farnum and Wilsman, 1983). Improved fixation and the use of ruthenium complexes has demonstrated the association of matrix components such as collagen and proteoglycan and the specific contact of pericellular proteoglycans with the chondrocyte plasmalemma (Shepard and Mitchell, 1977; Hunziker, Herrmann and Schenk, 1983; Hunziker and Schenk, 1984). These technical advances, along with the use of immuno-electron microscopy of specific matrix products (proteoglycans, collagen, link proteins) and electron spectroscopic imaging of elements such as calcium, phosphorus and sulphur (Poole *et al.*, 1982; Arsenault and Ottensmeyer, 1983), should enable us to break the code of cartilage's structural organisation within the near future.

The formation of joints

Murray dealt with the problem of the origin of the articulations which unite skeletal elements, both the freely movable synovial joints between long bones (pp. 7–26, 81–9) and sutures between the dermal bones of the cranium (pp. 65–7). He was concerned with the factors which determined where an articulation would develop, with initiation of the joint and with what moulds its final shape, and used evidence from pseudarthroses to show that joints could form in other than their normal sites and that both the initiation and the morphogenesis of such false joints were under mechanical control. A similar mechanism has been invoked by evolutionary biologists to explain the origin of new joints during evolution (Bock, 1960; Frazzetta, 1970). Murray showed that, in contrast, *primary* joints developed independently of any such mechanical stimulus, although he did not rule out 'developmental interactions between the skeletal segments' (p. 13) as playing a role. His chorioallantoic grafting experiments showed how normal the articular surfaces of such isolated elements could be (see his fig. 3, p. 15). Subsequent work has substantiated his views, a reasonable summary being that initiation of joints is independent of function while final differentiation and maintenance of the form of a joint are dependent upon extrinsic mechanical factors (see the papers in Sokoloff, 1978, 1980). Formation of the joint often only proceeds as far as initial cavitation, function being required to complete both cavitation and further differentiation, as well as to maintain the joint (Drachman and Coulombre, 1962; Drachman and Sokoloff, 1966; Beckham, Dimond and Greenlee, 1977; Mitrovic, 1982). Forces generated by growth of the adjacent skeletal element are not sufficient (Holder, 1977), although cell alignment is important, as it is for cartilage morphogenesis (Henrikson and Cohen, 1965). In a paper published after Murray's death (Murray and Drachman, 1969), paralysis was used to demonstrate how joints that are already formed ankylose when deprived of mechanical stimulation (see also Sullivan, 1966, 1974).

The development of sutures in the cranium has been a particularly difficult area of investigation. Are sutures predetermined boundaries or do they merely develop where two bones happen to meet? (p. 65). Along with this problem goes that of the amount of bone growth which occurs at the sutural margin. Classically, ostcogenesis at sutures was considered to provide a minimal contribution to growth of the skull (Brash, 1934). It is now known that, although growth can be quite different and specific, both on opposite faces of the bone and on opposite margins of the suture, such growth is adaptive and not a driving force in skull growth (Young, 1962; Moore and Lavelle, 1974; Moore, 1981). Furthermore, bone can be deposited where a suture would normally develop when sutural tissue is extirpated or when sutural fusion occurs – i.e. presumptive sutural soft tissue has osteogenic potential (Moss, 1959; Ritsila, Alhopuro and Ranta, 1973; and Hall, 1978, for a review).

The structure of sutures is especially well adapted to resist tension. Local areas of pressure or compression are met by the ability of cells in the suture to become chondrogenic, thus resisting and adapting to the local needs (Beresford, 1981). Again, intrinsic skeletogenic potential, expressed under extrinsic control, is paramount.

The trajectory theory and bone structure

Bone as an organ serves two primary functions, for it provides both a structural support and a reserve of calcium and phosphorus. As a *tissue*, bone is adapted to modify its behaviour and/or structure to serve these two functions, in the former case by growth or remodelling in response to mechanical stresses and in the latter by responding to altered serum levels of calcium and phosphorus and to changed hormonal levels.

A considerable portion of *Bones* is devoted to the trabecular structure of bone and to bone structure as adapted to and accurately mirroring the stress to which it is subjected (pp. 94–135, 155–74). The trajectory theory of Wolff posits that the trabeculae of cancellous bone follow the lines of trajectories in a homogeneous body of the same form as the bone and stressed in the same way. They are thus regarded as actually materialised trajectories (pp. 99–100 and see figs. 29–30, pp. 108–9). Murray reviewed several lines of evidence – the structure of the trabeculae in the long bones of marine mammals such as the dolphin (fig. 23, p. 100), fibre orientation, osteon arrangement, muscle action, the structure of the bone which forms in tendons, and the absence of tension trajectories, the temporal correlation between orientation of trabeculae and onset of function – and concluded 'that bony structures are therefore in a general way mechanically adaptive, but that the trajectorial theory in its rigid form demands a degree of perfection

in this adaptation which does not exist in fact, because the
mechanical stresses and strains of function are not the only
factors which influence the development of bony architectures'
(p. 135).

What is the current status of the trajectory theory and do we
now know how bones adapt their structure to meet an altered
mechanical environment? Although specialists such as ortho-
paedic and oral surgeons and orthodontists spend their working
lives manipulating the skeleton in conformity with the
trajectory theory, little has been explicitly done on it over the
past several decades. Is this because, like gravity which it
resists, it is no longer a theory but a truism, to be assumed and
therefore left unstated, or is it because it is a theory which is no
longer accepted, bone having to conform to various extrinsic
factors and not just to stress? Examination of numerous recent
texts reveals either acceptance of only a general moulding of
form along trajectorial lines (Urist, 1980, p. 114; Owen,
Goodfellow and Bullough, 1980, p. 68), arguments that a strict
trajectory model must be incorrect (Ashton and Holmes, 1981,
pp. 147–50), or, in most texts, no mention of the theory at all.

Perhaps the textbooks are lagging behind research in the
field. Several studies in the primary literature have examined
trabecular structure in the light of the trajectory theory.
Currey (1968) provides a very thoughtful review of some of
these studies.

Felts and Spurrell (1965, 1966) examined whale humeri but
could not provide strong support for the theory, structure
being related more to the physiological adaptations of life in
cold water. Dempster (1967) expressed similar caution from his
studies on the structure of the human skull, emphasising the
weakness of bone in tension, and that, at best, trabeculae
follow lines of maximum stress. Enlow (1968) emphasised the
importance of considering the structure and response of bone

to stress in the context of its soft tissue matrix (what Moss (1962) termed the functional matrix), a matrix which may dramatically modify the response of the bone when compared with how an isolated skeletal element might respond. This was also emphasised by Lanyon (1974) in his strain gauge analyses. The point is nicely illustrated by the biomechanics of the oral region of the dog (Scapino, 1967). In the upper jaw it is the anterior teeth on the maxilla which absorb much of the force generated in biting, while in the mandible the palatal mucosa develops a specialised, highly fibrous connective tissue which absorbs these forces; these two quite different adaptations to resist mechanical stress both result in a bony architecture considerably modified from that which would be predicted if the forces acted directly upon the bone of the jaws. Similar conclusions have been drawn from analysis of the avian bill (Bock, 1966). Nor does trabecular architecture correlate with mechanical forces in very small mammals, such as bats and shrews. Wolff's law does not apply, and instead intrinsic tissue strength moulds bony architecture in such animals (Dawson, 1980, and see also Hanken, 1982, for small salamanders). Similarly, in the development and structure of the vertebrae of bony fishes, the fundamental biconid shape is not stress-related but rather is a consequence of the vertebrae developing from mesenchyme which lies beside a notochord alternately constricted and dilated along the embryonic axis. Only the subsequently formed trabeculae of the spongy bone are deposited along lines dictated by mechanical stresses (Laerm, 1976). The final structure of the bones reflects both major influences on their development. On the other hand the *sequence* of ossification is functionally related, the first-formed bones being those associated with resisting the mechanical stresses of feeding and respiration (Weisel, 1967).

The trajectory theory presents idealised structures formed

in response to pure forces by an isolated bone whose only function is to resist those mechanical forces. Theoretical approaches which treat bones as idealised, isolated units, as in the application of catastrophe theory (Sinha, 1981), simply fall short of reality. A particular bone's response to altered mechanical stress might be compromised by the simultaneous response of the attached muscle or connective tissue (muscles may insert onto resorptive surfaces in which case muscular activity will stimulate bone resorption and not deposition – a factor not usually taken into account: Hoyte and Enlow, 1966; Enlow, 1975), by altered blood flow, by associated mineral requirements, etc. Just as in the embryonic condensations, so in the adult we cannot divorce the skeleton from its environment. Murray foresaw this interactive view and elegantly expressed it in his concluding chapter (pp. 177–80): 'Every bony structure is a compromise and no compromise is perfect' (p. 179). 'It is the question of the kind of functional stimulus which is effective and the manner of its action that remain as the most obscure problem in skeletogenesis' (pp. 179–80). Some of these stimuli will be discussed in the next section.

Growth of cartilage and bone

Murray considered several aspects of growth, including determination of the lengths of the skeletal elements, and concluded that 'undiscovered factors' such as those controlling rate of spread of chondrogenesis and relative position of chondrogenic centres must be important (p. 25). Given that the chondrogenic cells have to migrate to their final sites (see p. *xv*), relative position is determined by components of extracellular matrices (collagen, fibronectin, hyaluronate, etc.) which control migration (Newgreen, 1982*a*, *b*, 1984), and by the epigenetic factors which signal that migration should cease, the primary candidates being epithelia or nerves (Moss, 1972a; Hall, 1978; Horder, 1978; Lumsden, 1979). The spread of chondrogenesis is a function of the rate at which cells mature and deposit extracellular matrix-processes which are primarily controlled by hormones and by the physical constraint of the perichondrium (Wolpert, 1981; Hinchliffe and Johnson, 1983; Thorogood, 1983). Growth of cartilage is under intrinsic control by factors such as the number of stem cells in a condensation and the proportion of those cells capable of mitosis; the amount and rate of hypertrophy of chondrocytes; and the amount and rate of synthesis, secretion and degradation of extracellular matrix (Moss, 1972b; Hall, 1978; Hinchliffe and Johnson, 1983). The latter two (hypertrophy and secretion of extracellular matrix) are much more important than cell division, a finding that would perhaps not have been predicted but which is now very well documented (Biggers and Gwatkin, 1964; Hall, 1978; Thorogood, 1983). A particularly well-studied example is the development and differential growth of the tibia and fibula in the embryonic chick, where the original notion of competitive interaction for limited mesenchyme to form the initial condensations has been augmented by studies which focus on the rate of hypertrophy and matrix synthesis; these studies have been reviewed by Hinchliffe and Johnson (1983) and by Hall (1984). Hypertrophy and cell division are separately regulated. Thyroid hormone stimulates chondrocyte hypertrophy without affecting cell division (Burch and Lebovitz, 1982). Freezing at -80 °C suppresses growth by inhibiting hypertrophy without affecting cell division (Biggers, 1957).

There is considerable evidence for feedback from extracellular matrix to the chondrocytes as a factor in cartilage growth, both *in vivo* and *in vitro* (reviewed by Hall, 1978). The various genes known to produce achondroplasia (chondro-dystrophy) – genes such as *cartilage anomaly* (*can*), *achondro-plasia* (*cn*), *chondrodysplasia* (*cho*), *brachymorphic* (*bm*) and *stumpy* (*stm*) in the mouse, *achondroplasia* (*ac*) in the rabbit, and *nanomelia* (*nm*) in the chick – all produce a similar growth inhibition and syndrome but act on quite different processes affecting growth, viz., protein synthesis (*can*), chondrocyte maturation (*cn*), glycosaminoglycan synthesis (*cho*), aggrega-tion of glycosaminoglycans (*bm*), failure of chondroblasts to separate (*stm*), glucose metabolism (*ac*), or suppression of synthesis of cartilage-specific proteoglycan (*nm*) (Hall, 1978). These mutants provide a powerful example of the multiplicity of intrinsic controls over cartilage growth.

Hormones, vitamins and growth factors produced by cartilages themselves (cartilage-derived growth factor, chon-drocyte growth factor, cartilage-derived factor) have profound effects on the growth of cartilage (Elmer, 1983; Ornoy and Zusman, 1983; Silbermann, 1983).

Murray also considered the growth of the bone which replaces the cartilaginous model during endochondral ossifica-tion, both the increase in thickness and length (pp. 70–80) and the excess deposition which occurs on the concave surface of a curved bone (pp. 31–41 and see figs. 5–8, pp. 32–4). He interpreted growth in length and width as a response to functional demand, i.e. although growth of the cartilaginous model is largely under intrinsic control, growth of the bone which replaces it is largely under extrinsic control. This con-clusion has been amply confirmed by subsequent studies (see reviews by Moffett (1972), Goss (1978), Dixon and Sarnat (1982) and Hinchliffe and Johnson (1983), the last providing a systematic survey of the various groups of vertebrates). This extrinsic control is especially evident in the growth of the chondro- and osteocranium, where the sense organs and brain provide a functional matrix which regulates the rate of cartilage or bone growth. Skeleton, muscle, tendon and ligament act as a functional unit in controlling growth, a point emphasised by Murray (p. 91) and by many later studies (see Moss, 1972 *a*; Moffett, 1972; Moore and Lavelle, 1974; Moore, 1981, for discussions). Muscular control is classically illustrated by (*a*) the cessation of growth that is seen in mammalian mandibular processes such as the condylar or coronoid process when specific muscles are ablated or congenitally absent, or the increased growth which accompanies augmentation of muscle mass (Moore, 1965); (*b*) the effect of genes such as *muscular dysgenesis* which suppresses muscle development and therefore inhibits skeletal growth (Pai, 1965 *a, b*; Herring and Lakars, 1981); or (*c*) the effect of embryonic paralysis, prolonged bedrest or weightlessness on skeletal growth. The phenomenon of 'catch-up growth' further illustrates this extrinsic or epigenetic control (Williams, 1981). Extrinsic control of bone curvature has been elegantly demonstrated by Lanyon (1980) using unilateral sciatic neurectomy of rats. As he points out the altered tibial growth has to both accommodate altered muscular activity and optimise the new functional strains to which the tibia is exposed. Similar adaptations occur in bipedal rats (Riesenfeld, 1966).

Murray paid some attention to the differing effects of intermittent and constant pressure on bone growth (pp. 156 ff). Constant pressure results in atrophy while release from that pressure or alternating pressure and lack of pressure favours growth (pp. 156, 163). It is now known that the response of various cell types (myoblasts, fibroblasts, periosteal progenitor cells) to cyclic mechanical stimulation is increased DNA synthesis and cell division (Leung, Glagov and Mathews, 1976; 1977; Curtis and Seehar, 1978; Hall, 1979). The forces to which most skeletal elements are exposed derive from respiration, blood flow or locomotion, all of which produce cycles, often of very short duration, of alternating tension and compression (Rodbard, 1970; Liskova and Hert, 1971; Lanyon, 1972; Lanyon *et al.*, 1975; Goodship, Lanyon and McFie, 1979). Only some models take this intermittency of mechanical stress into account (Gjelsvik, 1973*a*), but these are the physiological stresses to which cartilage and bone respond. Weightlessness under zero gravity in space shows how different skeletal cells are adapted to respond differently to the same stimulus, for weightlessness slows bone formation but not resorption (Jee *et al.*, 1983), and weight-bearing bones are at greater risk than are non-weight-bearing bones (Simmons *et al.*, 1983). It is not only growth which is regulated by such stimuli. The pathway of cell differentiation can also be selected in response to the type of mechanical stimulus. Periosteal progenitor cells on membrane bones respond to intermittent stresses by forming secondary cartilage, i.e. by differentiating as chondroblasts, and to constant pressure by forming bone, i.e. by differentiating as osteoblasts (Murray, 1963; Hall, 1968, 1979). Chondrocytes, rather than fat cells (adipocytes) or connective tissue cells (fibroblasts), form in mesenchyme or the vascular system when these tissues are exposed to intermittent mechanical stress (Rodbard, 1970; Yamada *et al.*, 1977; Hadjiisky *et al.*, 1979; Arbustini, Jones and Ferrans, 1983). Chondrogenesis is not initiated when constant pressure is applied to bone (Glücksmann, 1938; Hall, 1968). In this we come full circle in Murray's book, for here we have determined cells responding to extrinsic signals to initiate or promote skeletogenesis.

The critical question now is not whether skeletal growth and structure are controlled both intrinsically and extrinsically, for *Bones* and the last half century have answered that in the affirmative, but how the two levels are coordinated to effect skeletogenesis. Mechanical effects on cartilage and bone may be mediated by altered electrical activity – for the skeleton is piezoelectric and both cartilage and bone respond to changes in electrical activity (Currey, 1968; Gjelsvik, 1973 *a*, *b*; Brighton, Black and Pollack, 1979; Hall, 1983 *c*) – or by changes in calcium and cyclic AMP levels (Bourret and Rodan, 1976; discussed in Hall, 1978). Mechanisms of action of hormones as gene regulators and of vitamins acting via the cell surface are beginning to be understood (Silbermann, 1983; Ornoy and Zusman, 1983). Many growth factors which act specifically on cartilage or on bone are now known (Elmer, 1983; Mohan *et al.*, 1984). We need more understanding of these factors, but the fundamental levels of control still elude us. What does determine the size and shape of a condensation, induce chondrocyte hypertrophy, attract osteoclast precursors to bone, set the balance between mechanical needs and ion reserve as elaborated in bone structure, determine the width of a suture, establish differential growth rates, produce a rod of cartilage and not a sheet? What is the fundamental structural unit of cartilage? As Murray said in concluding *Bones*, these are 'problems whose answers would throw a flood of light on the obscure processes of skeletogenesis'.

APPENDIX

*General references pertaining to the development, morphogenesis
and structure of the skeleton
An asterisk indicates a title which covers more than one topic.*

Cartilage

*BERESFORD, W. A. (1981). *Chondroid Bone, Secondary Cartilage and
Metaplasia.* Urban and Schwarzenberg, Baltimore and Munich.

FREEMAN, M. A. R. (ed.) (1973). *Adult Articular Cartilage.* Grune
and Stratton, New York.

*HALL, B. K. (1978). *Developmental and Cellular Skeletal Biology.*
Academic Press, New York.

HALL, B. K. (ed.) (1983). *Cartilage,* vols. 1–3. Academic Press,
New York.

*HAM, A. W. and CORMACK, D. H. (1978). *Histophysiology of
Cartilage, Bone and Joints.* J. B. Lippincott, Philadelphia.

KNESE, K.-H. (1979). *Stützgewebe und Skelettsystem.* Handbuch der
mikroskopische Anatomie der Menschen, vol. 2, part 5.
Springer-Verlag, Berlin and New York.

SERAFINI-FRACASSINI, A. and SMITH, J. W. (1974). *The Structure and
Biochemistry of Cartilage.* Churchill, London.

STOCKWELL, R. A. (1979). *Biology of Cartilage Cells.* Biological
Structure and Function 7. Cambridge University Press.

Bone

*BOURNE, G. H. (ed.) (1971, 1972, 1976). *The Biochemistry and
Physiology of Bone,* 2nd edn, vols. 1–4. Academic Press,
New York.

EL-NAJJAR, M. Y. and MCWILLIAMS, K. R. (1978). *Forensic
Anthropology. The Structure, Morphology, and Variation of Human
Bone and Dentition.* Charles C. Thomas, Springfield, Ill.

*HALSTEAD, L. B. (1974). *Vertebrate Hard Tissues.* Wykeham
Publications, London.

Hancox, H. M. (1972). *Biology of Bone.* Biological Structure and Function 1. Cambridge University Press.

*Kunin, A. S. and Simmons, D. J. (eds.) (1983). *Skeletal Research: An Experimental Approach*, vol. 2. Academic Press, New York.

Little, K. (1973). *Bone Behaviour.* Academic Press, London.

*McLean, F. C. and Urist, M. R. (1968). *Bone. Fundamentals of the Physiology of Skeletal Tissues*, 3rd edn. University of Chicago Press, Chicago.

*Peck, W. A. (ed.) (1982, 1983). *Bone and Mineral Research*, annuals 1 and 2. Elsevier, Amsterdam and New York.

*Simmons, D. J. and Kunin, A. S. (eds.) (1979). *Skeletal Research: An Experimental Approach*, vol. 1. Academic Press, New York.

*Urist, M. R. (ed.) (1980). *Fundamental and Clinical Bone Physiology.* J. B. Lippincott, Philadelphia.

*Vaughan, J. M. (1981). *The Physiology of Bone*, 3rd edn. Clarendon Press, Oxford.

Joints

Barnett, C. H., Davies, D. V. and MacConnill, M. A. (1961). *Synovial Joints: Their Structure and Mechanics.* Charles C. Thomas, Springfield, Ill.

Sokoloff, L. (ed.) (1978, 1980). *The Joints and Synovial Fluid*, vols. 1 and 2. Academic Press, New York.

Limbs and vertebrae

Ede, D. A., Hinchliffe, J. R. and Balls, M. (eds.) (1977). *Vertebrate Limb and Somite Morphogenesis.* British Society for Developmental Biology Symposium 3. Cambridge University Press.

*Fallon, J. F. and Caplan, A. I. (eds.) (1983). *Limb Development and Regeneration*, part A. Alan R. Liss, New York.

Hinchliffe, J. R. and Johnson, D. R. (1980). *The Development of the Vertebrate Limb: An Approach through Experiment, Genetics, and Evolution.* Clarendon Press, Oxford.

*Kelley, R. O., Goetinck, P. F. and MacCabe, J. A. (eds.) (1982). *Limb Development and Regeneration*, part B. Alan R. Liss, New York.

MILAIRE, J. (1974). *Histochemical Aspects of Organogenesis in Vertebrates*, part I, *The Skeletal System, Limb Morphogenesis, the Sense Organs*. Handbuch der Histochemie, vol. 8, part 3. Gustav Fischer Verlag, Stuttgart.

SWINYARD, C. A. (ed.) (1969). *Limb Development and Deformity: Problems of Evaluation and Rehabilitation*. Charles C. Thomas, Springfield, Ill.

Craniofacial skeleton

BERGSMA, D. (ed.) (1975). *Morphogenesis and Malformation of Face and Brain*. Alan R. Liss, New York.

BOSMA, J. F. (ed.) (1976). *Symposium on Development of the Basicranium*. DHEW Publication (NIH) 76–989. US Dept of Health, Education and Welfare, Bethesda, Md.

*GORLIN, R. J. (ed.) (1980). *Morphogenesis and Malformation of the Ear*. Alan R. Liss, New York.

MOORE, W. J. (1981). *The Mammalian Skull*. Biological Structure and Function 8. Cambridge University Press.

*PRATT, R. M. and CHRISTIANSEN, R. L. (eds.) (1980). *Current Research Trends in Prenatal Craniofacial Development*. Elsevier/ North-Holland, New York.

Growth

*DIXON, A. D. and SARNAT, B. G. (eds.) (1982). *Factors and Mechanisms Influencing Bone Growth*. Alan R. Liss, New York.

ENLOW, D. H. (1975). *Handbook of Facial Growth*. W. B. Saunders, Philadelphia.

GOOSE, D. H. and APPLETON, J. (1982). *Human Dentofacial Growth*. Pergamon Press, Oxford.

GOSS, R. J. (1978). *The Physiology of Growth*. Academic Press, New York.

MOORE, W. J. and LAVELLE, C. L. B. (1974). *Growth of the Facial Skeleton in the Hominoidea*. Academic Press, London.

Vascularity

BROOKES, M. (1971). *The Blood Supply of Bone: An Approach to Bone Biology*. Butterworth, London.

Ectopic skeletogenesis

CONNOR, J. M. (1983). *Soft Tissue Ossification*. Springer-Verlag, Berlin.
*FRIEDERSTEIN, A. Y. and LALYKINA, K. S. (1973). *Bone Induction and Osteogenic Precursor Cells*. Medical Publishing House, Moscow.

Mechanical stress

CURREY, J. D. (1970). *Animal Skeletons*. Edward Arnold, London.
CURREY, J. D. (1984). *The Mechanical Adaptations of Bones*. Princeton University Press, New Jersey.
EVANS, F. G. (1957). *Stress and Strain in Bone*. Charles C. Thomas, Springfield, Ill.
FROST, H. M. (1964). *The Laws of Bone Structure*. Charles C. Thomas, Springfield, Ill.
JAWORSKI, Z. F. G. (ed.) (1976). *Bone Morphometry*. University of Ottawa Press, Ottawa.
WAINWRIGHT, S. A., BIGGS, W. D., CURREY, J. D. and GOSLINE, J. M. (1976). *Mechanical Design in Organisms*. Edward Arnold, London.

Electrical factors

*BRIGHTON, C. T., BLACK, J. and POLLACK, S. R. (eds.) (1979). *Electrical Properties of Bone and Cartilage: Experimental Effects and Clinical Applications*. Grune and Stratton, New York.

LITERATURE CITED

ABBOTT, J. and HOLTZER, H. (1966). The loss of phenotypic traits by differentiated cells. III. The reversible behaviour of chondrocytes in primary culture. *J. Cell Biol.* **28**, 473–87.

ANDERSON, H. C. (1976). Osteogenetic epithelial–mesenchymal cell interactions. *Clin. Orthop. Rel. Res.* **119**, 211–24.

ARBUSTINI, E., JONES, M. and FERRANS, V. J. (1983). Formation of cartilage in bioprosthetic valves implanted in sheep – a morphologic study. *Am. J. Cardiol.* **52**, 632–6.

ARCHER, C. W., ROONEY, P. and WOLPERT, L. (1983). The early growth and morphogenesis of limb cartilage. In *Limb Development and Regeneration*, part A, ed. J. F. Fallon and A. I. Caplan, pp. 267–78. Alan R. Liss, New York.

ARSENAULT, A. L. and OTTENSMEYER, F. P. (1983). Quantitative spatial distributions of calcium, phosphorus and sulfur in calcifying epiphysis by high resolution electron spectroscopic imaging. *Proc. Nat. Acad. Sci. USA* **80**, 1322–6.

ASHTON, E. H. and HOLMES, R. L. (eds.) (1981). *Perspectives in Primate Biology.* Symposium of the Zoological Society of London 46. Academic Press, London.

BASSETT, C. A. L. and RUEDI, T. P. (1966). Transformation of fibrous tissue to bone *in vivo. Nature, Lond.* **209**, 988–9.

BECKHAM, C., DIMOND, R. and GREENLEE, T. K. (1977). The role of movement in the development of a digital flexor tendon. *Am. J. Anat.* **150**, 443–60.

BERESFORD, W. A. (1981). *Chondroid Bone, Secondary Cartilage and Metaplasia.* Urban and Schwarzenberg, Baltimore and Munich.

BIGGERS, J. D. (1957). The growth of cartilaginous embryonic chick bones after freezing. *Experientia* **13**, 483–9.

BIGGERS, J. D. and GWATKIN, R. B. L. (1964). Effect of X-rays on the morphogenesis of the embryonic chick tibiotarsus. *Nature, Lond.* **202**, 152–4.

BOCK, W. J. (1960). Secondary articulations of the avian mandible. *Auk* **77**, 19–55.

BOCK, W. J. (1966). An approach to the functional analysis of bill shape. *Auk* **83**, 10–51.

BOURRET, L. A. and RODAN, G. A. (1976). The role of calcium in the inhibition of cAMP accumulation in epiphyseal cartilage cells exposed to physiological pressure. *J. Cell Physiol.* **88**, 353–62.

BOWEN, I. D. and LOCKSHIN, R. A. (eds.) (1981). *Cell Death in Biology and Pathology.* Chapman and Hall, London.

BOYDE, A. and JONES, S. J. (1983). Scanning electron microscopy of cartilage. In *Cartilage,* ed. B. K. Hall, vol. 1, pp. 105–48. Academic Press, New York.

BRASH, J. (1934). Some problems in the growth and developmental mechanics of bone. *Edinburgh Med. J.* **41**, 305–19 and 363–86.

BRIGHTON, C. T., BLACK, J. and POLLACK, S. R. (eds.) (1979). *Electrical Properties of Bone and Cartilage: Experimental Effects and Clinical Applications.* Grune and Stratton, New York.

BURCH, W. M. and LEBOVITZ, H. E. (1982). Triiodothyroxine stimulation of *in vitro* growth and maturation of embryonic chick cartilage. *Endocrinology* **111**, 462–8.

CHAMBERS, T. J. (1982). Osteoblasts release osteoclasts from calcitonin-induced quiescence. *J. Cell Sci.* **57**, 247–60.

CHEVALLIER, A., KIENY, M. and MAUGER, A. (1977). Limb–somite relationship: origin of the limb musculature. *J. Embryol. Exp. Morphol.* **41**, 245–58.

CHRIST, B., JACOB, H. J. and JACOB, M. (1977). Experimental analysis of the origin of the wing musculature in avian embryos. *Anat. Embryol.* **159**, 171–86.

CURREY, J. D. (1968). The adaptation of bones to stress. *J. Theor. Biol.* **20**, 91–106.

CURTIS, A. S. G. and SEEHAR, C. M. (1978). The control of cell division by tension or diffusion. *Nature, Lond.* **274**, 52–3.

DAWSON, D. L. (1980). Functional interpretation of the radiographic anatomy of the femora of *Myotis lucifugus, Pipistrellus subflavus* and *Blarina brevicauda. Am. J. Anat.* **157**, 1–15.

DEMPSTER, W. T. (1967). Correlation of types of cortical grain structure with architectural features of the human skull. *Am. J. Anat.* **120**, 7–32.

DIXON, A. D. and SARNAT, B. G. (eds.) (1982). *Factors and Mechanisms Influencing Bone Growth.* Alan R. Liss, New York.

DRACHMAN, D. B. and COULOMBRE, A. J. (1962). Experimental

clubfoot and arthrogryposis multiplex congenita. *Lancet* **ii**, 523–6.

DRACHMAN, D. B. and SOKOLOFF, L. (1966). The role of movement in embryonic joint development. *Dev. Biol.* **4**, 401–20.

EDE, D. A. (1971). Control of form and pattern in the vertebrate limb. In *Control Mechanisms of Growth and Differentiation*, Symposium of the Society for Experimental Biology 25, pp. 235–54. Cambridge University Press.

EDE, D. A. (1983). Cellular condensation and chondrogenesis. In *Cartilage*, ed. B. K. Hall, vol. 2, pp. 143–86. Academic Press, New York.

EDE, D. A., HINCHLIFFE, J. R. and BALLS, M. (eds.) (1977). *Vertebrate Limb and Somite Morphogenesis*. Symposium of the British Society for Developmental Biology 3. Cambridge University Press.

EDE, D. A. and WILBY, O. K. (1981). Golgi orientation and cell behaviour in the developing pattern of chondrogenic condensations in chick limb bud mesenchyme. *Histochem. J.* **13**, 615–30.

EISENSTEIN, R., LARSSON, S.-E., SORGENTE, N. and KUETTNER, K. E. (1973). Collagen–proteoglycan relationships in epiphyseal cartilage. *Am. J. Pathol.* **73**, 443–56.

EISENSTEIN, R., SORGENTE, N. and KUETTNER, K. E. (1971). Organisation of extracellular matrix in epiphyseal growth plate. *Am. J. Pathol.* **65**, 515–34.

ELMER, W. A. (1983). Growth factors and cartilage. In *Cartilage*, ed. B. K. Hall, vol. 2, pp. 369–400. Academic Press, New York.

ENLOW, D. H. (1968). Wolff's law and the factor of architectonic circumstance. *Am. J. Orthodont.* **54**, 803–22.

ENLOW, D. H. (1975). *Handbook of Facial Growth*. W. B. Saunders, Philadelphia.

FARNUM, C. E. and WILSMAN, N. J. (1983). Pericellular matrix of growth plate chondrocytes: a study using postfixation with osmium ferrocyanide. *J. Histochem. Cytochem.* **31**, 765–75.

FELL, H. B. (1933). Chondrogenesis in cultures of endosteum. *Proc. R. Soc. Lond.* **112B**, 417–27.

FELL, H. B. (1969). The effect of environment of skeletal tissue in culture. *Embryologia* **10**, 181–205.

FELL, H. B. (1976). The development of organ culture. In

Organ Culture in Biomedical Research, ed. M. Balls and
M. Monnickendam, pp. 1–13. Cambridge University Press.

FELTS, W. J. L. (1961). *In vivo* implantation as a technique in
skeletal biology. *Int. Rev. Cytol.* **12**, 243–302.

FELTS, W. J. L. and SPURRELL, F. A. (1965). Structural
orientation and density in Cetacean humeri. *Am. J. Anat.* **116**,
171–204.

FELTS, W. J. L. and SPURRELL, F. A. (1966). Some structural and
developmental characteristics of Cetacean (Odontocete) radii.
A study of adaptive osteogenesis. *Am. J. Anat.* **118**, 103–34.

FRAZZETTA, T. H. (1970). From hopeful monster to bolyerine
snakes? *Am. Nat.* **104**, 55–72.

FRIEDENSTEIN, A. J. (1973). Determined and inducible osteogenic
precursor cells. In *Hard Tissue Growth, Repair and Remineralization*,
CIBA Foundation Symposium 11, pp. 170–85. Elsevier,
Amsterdam.

FRIEDENSTEIN, A. J. (1976). Precursor cells of mechanocytes.
Int. Rev. Cytol. **4**, 327–59.

FYFE, D. M. and HALL, B. K. (1983). The origin of the ecto-
mesenchymal condensations which precede the development of
the bony scleral ossicles in the eyes of embryonic chicks.
J. Embryol. Exp. Morphol. **73**, 69-86.

GJELSVIK, A. (1973*a*). Bone remodelling and piezoelectricity: I.
J. Biomech. **6**, 69–77.

GJELSVIK, A. (1973*b*). Bone remodelling and piezoelectricity: II.
J. Biomech. **6**, 187–93.

GLOWACKI, J., TREPMAN, E. and FOLKMAN, J. (1983). Cell shape
and phenotypic expression in chondrocytes. *Proc. Soc. Exp. Biol.
Med.* **172**, 93–8.

GLÜCKSMANN, A. (1938). Studies on bone mechanics *in vitro*. I.
Influence of pressure on orientation of structure. *Anat. Rec.* **72**,
97–113.

GOODSHIP, A. E., LANYON, L. E. and McFIE, H. (1979). Functional
adaptation of bone to increased stress. An experimental study.
J. Bone Jt. Surg. **61A**, 539–46.

GOSS, R. J. (1978). *The Physiology of Growth*. Academic Press,
New York.

GRÜNEBERG, H. (1963). *The Pathology of Development. A Study of
Inherited Skeletal Disorders in Animals*. Blackwell Scientific
Publications, Oxford.

GRÜNEBERG, H. and WICKRAMARATNE, G. A. (1974). A re-examination of two skeletal mutants of the mouse, *vestigial tail (vt)* and *congenital hydrocephalus (ch)*. *J. Embryol. Exp. Morphol.* **31**, 207–22.

HADJIISKY, P., DONEV, S., RENAIS, J. and SCEBAT, L. (1979). Cartilage and bone formation in arterial wall. I. Morphological and histochemical aspects. *Basic Res. Cardiol.* **74**, 649–62.

HALL, B. K. (1968). *In vitro* studies on the mechanical evocation of adventitious cartilage in the chick. *J. Exp. Zool.* **168**, 283–306.

HALL, B. K. (1970). Cellular differentiation in skeletal tissues. *Biol. Rev. Cambridge Philos. Soc.* **45**, 455–84.

HALL, B. K. (1975). The origin and fate of osteoclasts. *Anat. Rec.* **183**, 1–11.

HALL, B. K. (1977). Chondrogenesis of the somitic mesoderm. *Adv. Anat. Embryol. Cell Biol.* **53**(4), 1–50.

HALL, B. K. (1978). *Developmental and Cellular Skeletal Biology*. Academic Press, New York.

HALL, B. K. (1979). Selective proliferation and accumulation of chondroprogenitor cells as the mode of action of biomechanical factors during secondary chondrogenesis. *Teratology* **20**, 81–92.

HALL, B. K. (1982). The role of tissue interactions in the growth of bone. In *Factors and Mechanisms Influencing Bone Growth*, ed. A. D. Dixon and B. G. Sarnat, pp. 205–16. Alan R. Liss, New York.

HALL, B. K. (1983*a*). Tissue interactions and chondrogenesis. In *Cartilage*, ed. B. K. Hall, vol. 2, pp. 187–222. Academic Press, New York.

HALL, B. K. (1983*b*). Epithelial–mesenchymal interactions in cartilage and bone development. In *Epithelial-Mesenchymal Interactions in Development*, ed. R. H. Sawyer and J. F. Fallon, pp. 189–214. Praeger Press, New York.

HALL, B. K. (1983*c*). Bioelectricity and cartilage. In *Cartilage*, ed. B. K. Hall, vol. 3, pp. 309–38. Academic Press, New York.

HALL, B. K. (1984). Developmental mechanisms underlying the formation of atavisms. *Biol. Rev. Cambridge Philos. Soc.* **59**, 89-124.

HAM, A. W. and CORMACK, D. H. (1979). *Histophysiology of Cartilage, Bone and Joints*. J. B. Lippincott, Philadelphia.

HANKEN, J. (1982). Appendicular skeletal morphology in minute salamanders, genus *Thorius* (Amphibia: Plethodontidae): growth regulation, adult size determination, and natural variation. *J. Morphol.* **174**, 57–77.

HENRIKSON, R. C. and COHEN, A. S. (1965). Light and electron microscopic observations of the developing chick interphalangeal joint. *J. Ultrastruct. Res.* **13**, 129–62.

HERRING, S. W. and LAKARS, T. C. (1981). Craniofacial development in the absence of muscle contraction. *J. Craniofac. Gen. Dev. Biol.* **1**, 341–57.

HINCHLIFFE, J. R. (1977). The chondrogenic pattern in chick limb morphogenesis: a problem of development and evolution. In *Vertebrate Limb and Somite Morphogenesis*, ed. D. A. Ede, J. R. Hinchliffe and M. Balls, Symposium of the British Society for Developmental Biology 3, pp. 293–310. Cambridge University Press.

HINCHLIFFE, J. R. and JOHNSON, D. R. (1980). *The Development of the Vertebrate Limb.* Oxford University Press.

HINCHLIFFE, J. R. and JOHNSON, D. R. (1983). Growth of cartilage. In *Cartilage*, ed. B. K. Hall, vol. 2, pp. 255–96. Academic Press, New York.

HOLDER, N. (1977). An experimental investigation into the early development of the chick elbow joint. *J. Embryol. Exp. Morphol.* **39**, 115–27.

HOLTROP, M. E., COX, K. A. and GLOWACKI, J. (1982). Cells of the mononuclear phagocytic system resorb implanted bone matrix: a histologic and ultrastructural study. *Calcif. Tissue Int.* **34**, 488–94.

HORDER, T. J. (1978). Functional adaptability and morphogenetic opportunism, the only rules for limb development? *Zoon* **6**, 181–92.

HOYTE, D. A. N. and ENLOW, D. H. (1966). Wolff's law and the problem of muscle attachment on resorptive surfaces of bone. *J. Phys. Anthropol.* **24**, 205–14.

HUGGINS, C. B. (1931). The formation of bone under the influence of epithelium of the urinary tract. *Arch. Surg., Chicago* **22**, 377–408.

HUNZIKER, E. B., HERRMANN, W. and SCHENK, R. K. (1983). Ruthenium hexammine trichloride (RHT) – mediated interaction between plasmalemmal components and pericellular matrix proteoglycans is responsible for the preservation of chondrocytic plasma membranes *in situ* during cartilage fixation. *J. Histochem. Cytochem.* **31**, 717–27.

HUNZIKER, E. B. and SCHENK, R. K. (1984). Cartilage ultrastructure after high pressure freezing, freeze substitution and low temperature embedding. II. Intercellular matrix ultrastructure-preservation of proteoglycans in their native state. *J. Cell Biol.* **98**, 277–82.

ITEN, L. E. and MURPHY, D. J. (1980). Growth of quail wing skeletal element in a host chick wing. *Am. Zool.* **20**, 739.

JEE, W. S. S., WRONSKI, T. J., MOREY, E. R. and KIMMEL, D. B. (1983). Effects of space flight on trabecular bone in rats. *Am. J. Physiol.* **244**, R310–R314.

KASHIWA, H. K., LUCHTEL, D. L. and PARK, H. Z. (1975). Chondroitin sulfate and electron lucent bodies in the pericellular rim about unshrunken hypertrophied chondrocytes of chick long bone. *Anat. Rec.* **183**, 359-72.

KRUKOWSKI, M. and KAHN, A. J. (1982). Inductive specificity of mineralized bone matrix in ectopic osteoclast differentiation. *Calcif. Tissue Int.* **34**, 474-9.

LAERM, J. (1976). The development, function and design of amphicoelous vertebrae in teleost fishes. *Zool. J. Linn. Soc.* **58**, 237-54.

LANYON, L. E. (1972). The prospect of encouraging osteogenesis. *Vet. Annual* **13**, 126-9.

LANYON, L. E. (1974). Experimental support for the trajectorial theory of bone structure. *J. Bone Jt. Surg.* **56B**, 160-6.

LANYON, L. E. (1980). The influence of function on the development of bone curvature. An experimental study on the rat tibia. *J. Zool., Lond.* **192**, 457–66.

LANYON, L. E., HAMPSON, W. G. J., GOODSHIP, A. E. and SHAH, J. S. (1975). Bone deformation recorded *in vivo* from strain gauges attached to the human tibial shaft. *Acta Orthop. Scand.* **46**, 256–68.

LASH, J. W. and VASAN, N. S. (1983). Glycosaminoglycans of cartilage. In *Cartilage*, ed. B. K. Hall, vol. 1, pp. 215–52. Academic Press, New York.

LE DOUARIN, N. (1982). *The Neural Crest*. Development and Cell Biology 12. Cambridge University Press.

LEUNG, D. Y. M., GLAGOV, S. and MATHEWS, M. B. (1976).

Cyclic stretching stimulates synthesis of matrix components by arterial smooth muscle cells *in vitro*. *Science* **191**, 475–7.

LEUNG, D. Y. M., GLAGOV, S. and MATHEWS, M. B. (1977). A new *in vitro* system for studying cell response to mechanical stimulation. Different effects of cyclic stretching and agitation on smooth muscle cell biosynthesis. *Exp. Cell Res.* **109**, 285–98.

LISKOVA, M. and HERT, J. (1971). Reaction of bone to mechanical stimuli. II. Periosteal and endosteal reaction of tibial diaphysis in rabbit to intermittent loading. *Folia Morphol.* **19**, 301–17.

LIU, C.-C. and BAYLINK, D. J. (1984). Differential response in alveolar bone osteoclasts residing at two different bone sites. *Calcif. Tissue Int.* **36**, 182–8.

LUMSDEN, A. G. S. (1979). Pattern formation in the molar dentition of the mouse. *J. Biol. Buccale* **7**, 77–104.

McCLURE, J. (1983). The effect of diphosphonates on heterotopic ossification in regenerating Achilles tendon of the mouse. *J. Pathol.* **139**, 419–30.

MALONE, J. D., TEITELBAUM, S. L., GRIFFIN, G. L., SENIOR, R. M. and KAHN, A. J. (1982). Recruitment of osteoclast precursors by purified bone matrix constituents. *J. Cell Biol.* **92**, 227–30.

MARKS, S. C. Jr (1983). The origin of osteoclasts: evidence, clinical implications and investigative challenges of an extra-skeletal source. *J. Oral Pathol.* **12**, 226–56.

MAYNE, R. and VON DER MARK, K. (1983). Collagens of cartilage. In *Cartilage*, ed B. K. Hall, vol. 1, pp. 181–214. Academic Press, New York.

MITROVIC, D. (1982). Development of the articular cavity in paralyzed chick embryos and in chick embryo limb buds cultured on chorioallantoic membranes. *Acta Anat.* **113**, 313–24.

MOFFETT, B. C. (ed.) (1972). *Mechanisms and Regulation of Craniofacial Morphogenesis.* Swets and Zeitlander, Amsterdam.

MOHAN, S., LINKHART, T., FARLEY, J. and BAYLINK, D. (1984). Bone-derived factors active on bone cells. *Calcif. Tissue Int.* **36**, S139–S145.

MOORE, W. J. (1965). Masticatory function and skull growth. *J. Zool., Lond.* **146**, 123–31.

MOORE, W. J. (1981). *The Mammalian Skull.* Biological Structure and Function 8. Cambridge University Press.

MOORE, W. J. and LAVELLE, C. L. B. (1974). *Growth of the Facial Skeleton in the Hominoidea.* Academic Press, London.

Moss, M. L. (1959). The pathogenesis of premature cranial synostosia in man. *Acta Anat.* **37**, 351–70.

Moss, M. L. (1962). The functional matrix. In *Vistas in Orthodontics*, ed. B. S. Kraus and R. A. Riedel, pp. 85–98. Lea and Febiger, Philadelphia.

Moss, M. L. (1972a). An introduction to the neurobiology of orofacial growth. *Acta Biotheor., Leiden* **22 A**, 236–59.

Moss, M. L. (1972b). The regulation of skeletal growth. In *Regulation of Organ and Tissue Growth*, ed. R. J. Goss, pp. 127–42. Academic Press, New York.

MUNDY, G. R., RODAN, S. B., MAJESKA, R. J., DeMARTINO, S., TRIMMIER, C., MARTIN, T. J. and RODAN, G. A. (1982). Unidirectional migration of osteosarcoma cells with osteoblast characteristics in response to products of bone resorption. *Calcif. Tissue Int.* **34**, 542–6.

MURRAY, P. D. F. (1957). Cartilage and bone – a problem in tissue differentiation. *Aust. J. Sci.* **14**, 65–73.

MURRAY, P. D. F. (1963). Adventitious (secondary) cartilage in the chick embryo and the development of certain bones and articulations in the chick skull. *Aust. J. Zool.* **11**, 368–430.

MURRAY, P. D. F. and DRACHMAN, D. B. (1969). The role of movement in the development of joints and related structures: the head and neck in the chick embryo. *J. Embryol. Exp. Morphol.* **22**, 349–71.

MURRAY, P. D. F. and SMILES, M. (1965). Factors in the evocation of adventitious (secondary) cartilage in the chick embryo. *Aust. J. Zool.* **13**, 351–81.

NEWGREEN, D. F. (1982a). Adhesion to extracellular materials by neural crest cells at the stage of initial migration. *Cell Tissue Res.* **227**, 297–318.

NEWGREEN, D. F. (1982b). The role of the extracellular matrix in the control of neural crest cell migration. In *Extracellular Matrix*, ed. S. Hawkes and J. L. Wang, pp. 141–6. Academic Press, New York.

NEWGREEN, D. F. (1984). Spreading of explants of embryonic chick mesenchymes and epithelia on fibronectin and laminin. *Cell Tissue Res.* **236**, 275–77.

NODEN, D. M. (1980). The migration and cytodifferentiation of cranial neural crest cells. In *Current Research Trends in Prenatal*

Craniofacial Development, ed. R. M. Pratt and R. L. Christiansen, pp. 3–26. Elsevier/North-Holland, New York.

NODEN, D. M. (1983). The role of the neural crest in patterning of avian cranial skeletal, connective, and muscle tissues. *Dev. Biol.* **96**, 144–65.

ORNOY, A. and ZUSMAN, I. (1983). Vitamins and cartilage. In *Cartilage*, ed. B. K. Hall, vol. 2, pp. 297–326. Academic Press, New York.

OSTER, G. F., MURRAY, J. D. and HARRIS, A. K. (1983). Mechanical aspects of mesenchymal morphogenesis. *J. Embryol. Exp. Morphol.* **78**, 83–125.

OWEN, M. (1970). The origin of bone cells. *Int. Rev. Cytol.* **28**, 213–38.

OWEN, M. (1978). Histogenesis of bone cells. *Calcif. Tissue Res.* **25**, 205–7.

OWEN, R., GOODFELLOW, J. and BULLOUGH, O. (eds.) (1980). *Scientific Foundation of Orthopaedics and Traumatology.* Heinemann Medical, London.

PAI, A. C. (1965*a*). Developmental genetics of a lethal mutation, *muscular dysgenesis (mdg)*, in the mouse. I. Genetic analysis and gross morphology. *Dev. Biol.* **11**, 82–92.

PAI, A. C. (1965*b*). Developmental genetics of a lethal mutation, *muscular dysgenesis (mdg)*, in the mouse. II. Developmental analysis. *Dev. Biol.* **11**, 93–109.

PARFITT, A. M. (1984). The cellular basis of bone remodelling: the quantum concept reexamined in light of recent advances in the cell biology of bone. *Calcif. Tissue Int.* **36**, S37–S45.

POOLE, A. R., PIDOUX, I., REINER, A. and ROSENBERG, L. (1982). An immunoelectron microscope study of the organization of proteoglycan monomer, link protein and collagen in the matrix of articular cartilage. *J. Cell Biol.* **93**, 921–37.

REDDI, A. H. (1983). Role of extracellular matrix in cell differentiation and morphogenesis. In *Epithelial–Mesenchymal Interactions in Development*, ed. R. H. Sawyer and J. F. Fallon, pp. 75–92. Praeger Press, New York.

RIESENFELD, A. (1966). The effects of experimental bipedalism and upright posture in the rat and their significance for the study of human evolution. *Acta Anat.* **65**, 449–521.

RITSILA, V., ALHOPURO, S. and RANTA, R. (1973). The role of the zygomatic arch in the growth of the skull in rabbits. *Proc. Finn. Dental Soc.* **69**, 164–5.

RODAN, G. A. and MARTIN, T. J. (1981). Role of osteoblasts in hormonal control of bone resorption – a hypothesis. *Calcif. Tissue Int.* **33**, 349–52.

RODBARD, S. (1970). Negative feedback mechanisms in the architecture and function of the connective and cardiovascular tissues. *Persp. Biol. Med.* **13**, 507–27.

ROGERS, W. P. (1968). Patrick Desmond Fitzgerald Murray. *Rec. Aust. Acad. Sci.* **1**, 71–6.

SCAPINO, R. P. (1967). Biomechanics of prehensile oral mucosa. *J. Morphol.* **122**, 89–114.

SHEPARD, N. and MITCHELL, N. (1977). The use of ruthenium red and *p*-phenylenediamine to stain cartilage simultaneously for light and electron microscopy. *J. Histochem. Cytochem.* **25**, 1163–8.

SILBERMANN, M. (1983). Hormones and cartilage. In *Cartilage*, ed. B. K. Hall, vol. 2, pp. 327–68. Academic Press, New York.

SIMMONS, D. J., RUSSELL, J. E., WINTER, F., TRAN VAN, P., VIGNERY, A., BARON, R., ROSENBERG, G. D. and WALKER, W. V. (1983). Effect of spaceflight on the non-weight-bearing bones of rat skeleton. *Am. J. Physiol.* **244**, R319–R326.

SINHA, D. K. (1981). Biomechanical problems and catastrophe theory. In *Catastrophe Theory and Applications*, ed. D. K. Sinha, pp. 100–4. John Wiley, New York.

SOKOLOFF, L. (1969). *The Biology of Degenerative Joint Disease.* University of Chicago Press, Chicago.

SOKOLOFF, L. (ed.) (1978). *Joints and Synovial Fluid*, vol. 1. Academic Press, New York.

SOKOLOFF, L. (ed.) (1980). *Joints and Synovial Fluid*, vol. 2. Academic Press, New York.

SOLURSH, M. (1983). Cell–cell interactions and chondrogenesis. In *Cartilage*, ed. B. K. Hall, vol. 2, pp. 121–42. Academic Press, New York.

STOCKWELL, R. A. (1979). *Biology of Cartilage Cells.* Biological Structure and Function 7. Cambridge University Press.

SULLIVAN, G. E. (1966). Prolonged paralysis of the chick embryo, with special reference to effects on the vertebral column. *Aust. J. Zool.* **14**, 1–17.

SULLIVAN, G. E. (1974). Skeletal abnormalities in chick embryos paralysed with decamethonium. *Aust. J. Zool.* **22**, 429–38.

THOROGOOD, P. V. (1983). Morphogenesis of cartilage. In *Cartilage*, ed. B. K. Hall, vol. 2, pp. 223–54. Academic Press, New York.

TRELSTAD, R. L. (1977). Mesenchymal cell polarity and morphogenesis of chick cartilage. *Dev. Biol.* **59**, 153–63.

TURING, A. M. (1952). The chemical basis of morphogenesis. *Phil. Trans. R. Soc. Lond.* **237B**, 37–72.

TWITTY, V. C. and SCHWIND, J. L. (1931). The growth of eyes and limbs transplanted heteroplastically between two species of *Amblystoma*. *J. Exp. Zool.* **59**, 61–86.

URIST, M. R. (ed.) (1980). *Fundamental and Clinical Bone Physiology*. J. B. Lippincott, Philadelphia.

URIST, M. R. (1983). The origin of cartilage: investigations in quest of chondrogenic DNA. In *Cartilage*, ed. B. K. Hall, vol. 2, pp. 2–86. Academic Press, New York.

URIST, M. R., DOWELL, T. A. and HAY, P. H. (1968). Bone induction and osteogenetic competence. *Am. Zool.* **8**, 783.

WEISEL, G. F. (1967). Early ossification in the skeleton of the sucker (*Catastomus macrocheilus*) and the guppy (*Poecilia reticulata*). *J. Morphol.* **121**, 1–18.

WEISS, P. and MOSCONA, A. (1958). Type-specific morphogenesis of cartilages developed from dissociated limb and scleral mesenchyme *in vitro*. *J. Embryol. Exp. Morphol.* **6**, 238–46.

WILLIAMS, J. P. G. (1981). Catch-up growth. *J. Embryol. Exp. Morphol.* **65** [Suppl.], 89–101.

WLODARSKI, K. (1969). The inductive properties of epithelial established cell lines. *Exp. Cell Res.* **57**, 446–8.

WOLPERT, L. (1978). Cell position and cell lineage in pattern formation and regulation. In *Stem Cells and Tissue Homeostasis*, ed. B. I. Lord., C. S. Potten and R. J. Cole, Symposium of the British Society for Cell Biology 2, pp. 29–47. Cambridge University Press.

WOLPERT, L. (1981). Cellular basis of skeletal growth during development. *Br. Med. Bull.* **37**, 215-19.

WOLPERT, L. and HORNBRUCH, A. (1981). Positional signalling along anteroposterior axis of the chick wing. The effect of multiple polarizing region grafts. *J. Embryol. Exp. Morphol.* **63**, 145–59.

WONG, Y. C. and BUCK, R. C. (1972). The development of sites of metaplastic change in regenerating tendon. *Z. Zellforsch. mikrosk. Anat.* **134**, 175–82.

YAMADA, M., FUJIMORE, K., TAKEUCHI, H., YAMAMOTO, K. and TAKAKUSU, A. (1977). Hematopoiesis in bovine heart bone. *Cell Struct. Function* **2**, 353–60.

YOUNG, R. W. (1962). Autoradiographic studies on postnatal growth of the skull in young rats injected with tritiated thymidine. *Anat. Rec.* **143**, 1–14.

ZAMBONIN ZALLONE, A., TETI, A. and PRIMAVERA, M. V. (1984). Resorptions of vital or devitalized bone by isolated osteoclasts *in vitro*. The role of lining cells. *Cell Tissue Res.* **235**, 561–64.

CHAPTER I

THE PRIMARY DEVELOPMENT OF THE SKELETON

The development of replacing bones falls into two parts: the preparation of the cartilaginous model and the development of the bony structure for which it provides a scaffolding. This chapter deals with the first phase, the development of the bony skeleton being left to chapter II. There is one exception to this arrangement: for the sake of convenience, and because of its possible causal relationship with the structural arrangement of the first trabeculae of bone, consideration of the manner in which the architecture of the cartilaginous elements is produced in the mammalian limb is postponed to the next chapter.

Nearly all the experimental studies of the primary development of replacing bones have been made on the long bones of the limb, and for this reason the discussion which follows is confined almost entirely to them.

The first visibly detectable rudiment of the skeleton is an axial condensation of the mesoderm in the limb-bud. The cells in this condensation are the precursors of those which later chondrify. The problem arises whether these are cells which, already predetermined for chondrification, have taken up an axial position, or indifferent cells which happen to be axially placed and which become chondrogenic because of the action of factors which are confined to this region of the limb-bud.

In an experiment which has not been published, but which I am kindly allowed to mention here, H. B. Fell cultivated *in vitro* tiny fragments of the teased mesenchyme of limb-buds of the four-day chick embryo. Cartilage developed in some fragments but not in others and chondrification was not correlated with the size of the fragments. It occurred in some explants of all sizes, down to the smallest, which consisted of only about thirty cells, but was absent even in some of the larger pieces, containing thousands of cells. It can be concluded from this that: (*a*) chondrification was not dependent either upon axial position of the cells in a large mass of tissue or upon any condition which could not exist in very tiny fragments of the limb-bud; (*b*) since all the explants were treated in the same way, and were lying side by side in the same medium, the differences between them must have been of intrinsic origin, and were not inflicted upon them by extrinsic factors. This is equivalent to saying that the chondrifying explants contained tissue predetermined for chondrification. In another experiment, also not yet published, and done in collaboration by W. Jacobson and H. B. Fell, and which again I am kindly allowed to mention, it has been found that the cells which form cartilage, bone, and muscle in the lower jaw of the chick embryo are derived from three distinct places of origin in the jaw, and then migrate to the positions in which their histogenic differentiation later occurs. When these regions are cultivated separately, one forms cartilage, another bone, and the third muscle.

In the face of these facts, the conclusion is irresistible that the cells of the axial condensation in such a structure as a limb-bud are already determined, or at least biased, towards the formation of cartilage and bone.

It should be noted that, within the limits of the skeletal tissues, there is, in both embryonic and adult life, considerable lability. Thus Fell (1933) found that under certain conditions of *in vitro* cultivation osteoblasts may become cartilage cells enclosed in capsules of matrix; it is interesting to note that this cartilaginous condition is unstable and in the majority of cases the cartilage becomes directly transformed into bone. Studitsky (1934*a*) found that the periosteum of the chick embryo, carefully taken only from the middle part of long bones, where no more cartilage would normally be formed, could produce cartilage when grafted into the chorio-allantois of older chicks. In adult mammals, the development of cartilage in bony callus is a common observation and there is often a cartilage covering over the articular surfaces of pseudarthroses. In the healing of fractures in adult birds, Roggemann (1930) observed that periosteal proliferation produced a blastema which became converted into cartilage, while the endosteum gave rise directly to spongy bone. The periosteal cartilage, becoming calcified, was subsequently resorbed and replaced by endochondral bone.

These changes are all within the group of skeleton-forming tissues, but it is a well-founded belief of pathologists that both cartilage and bone can, under conditions which remain obscure, be produced by "indifferent" cells of connective tissue which would not normally have differentiated in this manner. There can be little doubt that ectopic ossification, such as is seen clinically and has been produced experimentally by Huggins (1931) and others, is of this nature. Carey, Zeit and McGrath (1927–28) extirpated the patellae of puppies and found that regeneration occurred if movement of the knee-joint were permitted but not if it were prevented, and they drew the conclusion that cartilage is formed from indifferent cells in response to external conditions of a mechanical nature; that such conditions can call forth the differentiation of cartilage is demonstrated by the cartilage-covered articular surfaces of the more highly developed pseudarthroses. The experiments of Levi (1930), in which cartilage formation was experimentally induced in the region of the phalanges of guinea-pigs and chicks, point in the same direction. Niven (1933), however, selecting for her study the same cartilage, in another animal, as had been investigated by Carey and his colleagues, found that, when the mesenchyme which would later form the patella was extirpated from chick embryos at seven, eight, and nine days and cultivated *in vitro*, cartilage appeared in the explants, and in those from nine-day and some eight-day embryos it had the general form of the normal patella. Thus, while certain mechanical conditions are required for the regeneration of the patella from tissue which would not normally form cartilage (Carey and colleagues), the first formation of the patella is a self-differentiation, and neither functional activity nor mechanical forces due to growth pressure, etc. are required. It is therefore clear that the conclusions justly formed by Carey in regard to cartilage formation in post-embryonic life cannot be extended to the primary formation of the same tissue, even in the same organ, in embryonic life. Thus the formation under certain conditions of cartilage and bone by previously indifferent cells in no way invalidates the evidence that the cells which normally form these tissues in embryonic differentiation are predetermined at some time before the process actually begins.

The problem of the development of form above the histological level resolves itself into several parts which, since it is mainly the limb skeleton that is our present concern, are determined by the characteristics of that skeleton. These characteristics are: the shaft form of the bones, the different lengths of individual bones, the existence of joints between bones, the form of the bones and especially of the articular structures at their ends.

The shaft form is already indicated in the mesenchymal condensation; thus the first stages of its differentiation must occur at this time or earlier. A small amount of evidence indicates that it is earlier, but not with certainty. Chorio-allantoic grafts of fragments taken from the sides of two-day embryos, in the region in which the limb-buds will later appear, produce cartilaginous structures which have the shaft form, are jointed, and are branching (Murray, 1926, 1928). This certainly shows that the graft as a whole contains all the factors necessary for the formation of these structures, and therefore either (1) that the group of cells which was presumptive for skeleton formation was already intrinsically determined for shaft formation, or that (2) this was not so, but that factors acting in the graft but extrinsic to the presumptive skeletal cells caused the latter to assume the shaft arrangement. Now, the tissues present in the grafts were: ectoderm, mesenchyme presumptive for dermis formation, mesenchyme presumptive for skeleton formation, endoderm, and mesenchyme presumptive for smooth muscle formation. The first two formed skin, usually in vesicles, with dermis and feather

germs, the third formed the jointed and branched cartilaginous structures, the fourth and fifth formed sections of
intestine with smooth muscle; striped muscles were absent
in most. There was no trace of any normal or regular
anatomical relationship between the three sets of structures, so that the inter-relations between parts of the
grafts were effectively destroyed; it must be remembered
that these grafts were no more than little scraps of tissue
from the area pellucida of two-day embryos. Thus whatever arrangements of cells were brought into existence in
the skeletogenous mesenchyme were probably due to
factors operating within it; and this is another way of
stating the first alternative. It therefore seems likely that
there already exists in the skeletogenous mesenchyme,
before the limb-buds appear, some kind of plan, or growth
pattern, which is expressed in the formation of the axial
condensation in the limb-bud, where the shaft form is first
visibly foreshadowed.

The development of the curvatures of the shafts might
be caused either by the intrinsic growth pattern of the
element, by such extrinsic factors as muscular tension, or by
the tension of other soft parts relatively retarded in growth
(Carey, 1922). There can be no doubt that unbalanced
mechanical factors of this kind, especially if in excess of
those normally acting, would bring about curvature of the
shafts, but it is also certain that factors acting within the
developing shaft itself cause it to take on the correct
curvature (Murray, 1926; Murray and Selby, 1930). This
is shown by chorio-allantoic grafts of parts of four- and
five-day limb-buds and by grafts of isolated femora of six-
day embryos. In the grafts of fragments of limb-buds the

influence of such factors as the weight of the embryo was excluded, and the action of muscular pull practically excluded, since muscles either did not develop in the grafts at all or were quite irregular in their arrangement and usually exiguous in quantity. Nevertheless, there was a strong tendency for femoral shafts to show both the normal curvatures (fig. 2, p. 15). The same was true of grafts of isolated femora of six-day embryos, from which the developing musculature had been removed before grafting (fig. 4, p. 18). The same series of grafts showed many shafts, of various elements, which were bent (fig. 5, p. 32) either in abnormal directions or else to an abnormal extent in the normal directions. These bends were presumably the result of mechanical pressures acting upon the graft in the abnormal environment, pressures whose intensity was sufficiently great to overcome the intrinsic tendency to develop with the normal curvatures. It may, therefore, be concluded that the intrinsic factors which direct the growth of the shaft tend to cause it to become curved in the normal manner but that extrinsic mechanical factors may override the intrinsic factors if they become sufficiently strong. It is probable, of course, that the extrinsic factors which act in the developing limb are such as to favour the development of the normal form.

The other features in which the form of the elements is expressed are the shape of the articular structures and of such projections as trochanters.

So long as no method had been found whereby it was possible to study the behaviour of the half of a developing joint, isolated from its fellow, discussion of the development of joint forms centred round indirect evidence. Since the problem has been entirely transformed by the introduction of experimental methods, it is not necessary to discuss the older literature in detail. Varying degrees of importance were attached to three factors: the mechanical rubbing or grinding of opposed surfaces upon one another when the joints moved under the action of muscles, the mutual effects of the growth pressures of opposing elements, and "heredity" or self-differentiation. Ludwig Fick (1859), who opened the discussion, recognised the efficacy of both of the first two factors, considering that a convex articular surface developed on that element of a pair which had the higher growth intensity during the period of grinding, but Henke and Reyher (1874) denied that growth pressures played any part. Bernays (1878) produced histological evidence from which he concluded that joint form developed before there was any question of muscular activity; he found that the form of the joint surfaces was developed before the appearance of the joint cavity and before the development of functional muscles. The objections raised by Bernays can only doubtfully be upheld, for joints exist which are normally united by connective tissue but which nevertheless function apparently to the satisfaction of their owners, and embryonic muscular tissue can carry out contractions surprisingly early. Schulin (1879) agreed that muscles are not responsible for the formation of articular surfaces, but considered that their action produced the cavity, the articular surfaces being produced by opposed growth pressures.

R. Fick (1890, 1921, 1928) rubbed gypsum blocks against one another and found that attrition between them caused a change in the form of the rubbed surfaces. The lower block was held stationary while the upper block moved upon it. The new forms thus produced depended upon whether the force applied to the moving block was above or below its centre of gravity. If the force applied to the moving block acted upon it below its centre of gravity, the moving block became concave and the lower one convex (fig. 1 A), but if the force were applied above the centre of gravity it became convex and the lower block

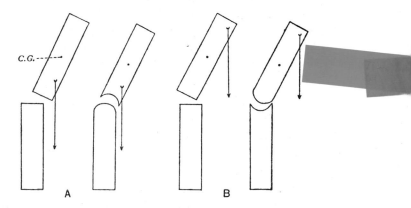

Fig. 1. *C.G.*=centre of gravity. (R. Fick, 1921.)

concave (fig. 1 B). Fick's description is given without reference to the centre of gravity: if the force acted on the moving block near its point of contact with the lower one, the rubbed end of the moving block became concave, but if remote from its contact with the lower block, convex. A study of the joints of Man and animals then showed (not without exception, Lubosch, 1910) a corresponding correlation between the forms of articular surfaces and the positions of action of muscles on the bone. Fick did not claim that the form of the joint surfaces was directly determined in the individual life-history by the mechanical conditions resulting from the positions of muscular attachment, but was content to point out that the form of joints corresponded with the mechanical rule and was therefore fit for survival through natural selection. He specifically rejected the view that joint forms are produced by direct rubbing between opposed surfaces, and adduced a number of reasons. The most convincing of these was that in certain cases there could not be sufficient movement of the embryonic limbs to enable all parts of the joint surface to be subjected to rubbing. He considers, for example, that in the case of the shoulder and knee-joints of birds and reptiles, the size of the egg renders impossible a sufficient freedom of movement for the whole of the opposed surfaces to move upon another. The entire head of the femur could not be formed by movements of the femur against the pelvis. Fick therefore considered that at least parts of the joint surfaces must develop by "reine Vererbung". His general conclusion is that, while heredity plays an important part in the development of joint form, this is nevertheless dependent on mechanical conditions and is subject to alteration with change of function. In particular, if the position of the muscular attachments be altered, the form of the joint may be reversed; this statement is supported by the experiment of Wachter, mentioned on p. 81.

It has become increasingly clear that foetal or embryonic movement cannot be regarded as the decisive factor in the primary development of joint form. Thus Hesser (1925) considers that the presence of undamaged mesenchyme between developing articular surfaces means that the joint cannot yet be capable of movement. He is thus in agreement with Bernays. The existence of normal and functioning joints which are nevertheless joined by an intermediate tissue detracts, of course, from the value of this argument, but the difficulty hardly applies to a particular case brought forward by Hesser. In the 27 mm. human embryo the eardrum is thick and cannot vibrate, hence the ear ossicles can hardly be functional and are indeed partly embedded in mesenchyme. Nevertheless the articulations have already their typical form. Hesser, of course, agrees with practically all other authors in recognising the dependency of normal later development upon the supervention of function.

Before leaving this phase of the controversy the work of Nauck (1926) must be mentioned. Dissatisfied with the theories of function, attrition, and growth pressure, he introduces the new conception of "restriction of growth" ("Wachstums- or Ausdehnungshemmung"). Examining the growth-pressure theory, and studying the development of a large number of joints from many groups, he accepts as a criterion of mechanical hardness the histological condition of the elements. A cartilaginous structure in a histologically advanced condition is doubtless tougher than one whose development is in an earlier stage. Now, were the growth-pressure theory correct, one would expect that, at the time when the joint surfaces are being modelled, the element which becomes convex would be harder or histologically more advanced than that which is to be concave. In the fore-limb of the pig the cartilage of the humerus head and pectoral girdle (in the relevant areas) develop equally rapidly, while in Man and *Triton* (newt) that of the humerus develops, as required by the theory, more rapidly than that of the girdle. Much the same applies to *Acanthias* (dog-fish). But in *Anas* (duck) the opposite condition is found, the cartilage of the girdle being ahead of that of the humerus, though the latter bears the convexity, the former the concavity. Further, the forearm of Man chondrifies before the carpus, but bears the concavity. Faced with these difficulties, Nauck discards the growth-pressure theory and substitutes for it his new conception of "limitation of extension". Taking as an example the development of the shoulder joint, Nauck considers the stage at which the coracoid and scapular cartilages are meeting. Between them lies the head of the humerus. The more differentiated parts of the humerus prevent the growth of the girdle elements in a distal direction; at the same time the cartilage of the head of the humerus, irrespective of whether it is "softer" or "harder" than that of the girdle elements, prevents the extension of these into the region which it occupies. At the same time the girdle elements surround and enclose the head of the humerus as in a cup, so preventing its further expansion. The explanation of two-component joints is along similar lines.

The first contribution in which experimentation played an important part was made by Braus (1909, 1910). He studied the normal development of the fore limb of the toad *Bombinator* and made ectopic grafts of the limb-buds. He found evidence which led him to conclude that the development of normal articular structures was not dependent either on the functional activity of muscles or on inter-relations between adjacent elements. This evidence was derived from both normal and grafted limbs. In normal development he found that the limbs neither showed spontaneous movement nor responded to electrical stimulation until about the time of metamorphosis, and he cites (1909) a larva, fixed while both spontaneous and stimulated movements were still absent, which nevertheless had a normal humeral head and glenoid cavity. In association with grafts there was sometimes formed, in addition to the primary limb, an accessory one, and these accessory limbs were always devoid of nerves and therefore paralysed. Their shoulder joints were nevertheless normally formed. The facts so far mentioned evidently exclude muscular activity as an essential factor in the development of articular structures, but do not exclude developmental interactions between the skeletal segments. In the grafts themselves, for reasons into which there is no need to go, it frequently happened that the shoulder girdle was subnormal in size but the free part of the limb of normal dimensions. There was found in such cases a large humeral head in association with a small glenoid cavity, so that the one did not fit the other. Each was normally developed, although, owing to the disparity in their sizes, the form of neither could have been produced by any mechanical interaction with the other.

The arguments of Braus, convincing though they may seem to us to-day, did not meet with general acceptance. This was probably inevitable, and indeed the proof was not and could not be complete until a method was found whereby the development of one component of a joint could be studied in the absence of the other component and of function.

In 1925 J. S. Huxley and I published a brief paper describing a femur (fig. 2) which was developed in a grafted basal fragment of the posterior limb-bud of a four-day chick embryo. The graft was made upon the chorio-allantois of an older embryo. When, after some days, the graft was recovered, fixed and sectioned, it was found to consist of a femur without any other skeletal parts except one small ectopic piece of cartilage. At the distal end condyles were little more than indicated but at the proximal end there was a well-developed head. The pelvis was entirely absent. A practically normal articular structure had thus been developed, in the absence of the other component of the joint and of all movement, in a graft which at the time of operation can have contained no cartilage. The work was continued and in 1926 I described another femur having a well-developed head, in the absence of the pelvis, a femur having condyles in the absence of tibia and fibula, and a humerus (fig. 3) having condyles in the absence of radius and ulna. It was also found that each grafted fragment of a limb-bud, with a few doubtful exceptions, produced almost exactly what it would have formed in normal development as part of an entire limb. The limb-bud at four days (probably also at three and perhaps at two days) is thus already a mosaic, with each region already determined for the development of some particular structure.

These facts are in themselves sufficient to eliminate as decisive in the development of articular structures of this kind, all such factors as movement, attrition, growth pressure, "Ausdehnungshemmung" and the like. In agreement with these conclusions are the experiments of Hamburger (1928), who found that complete denervation, and therefore complete paralysis, did not prevent the practically normal development of the limb skeletons of frogs. In six

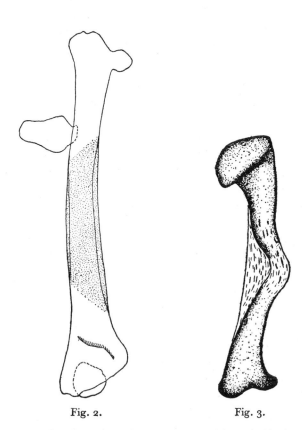

Fig. 2.   Fig. 3.

Fig. 2. Femur developed in a chorio-allantoic graft from the basal fragment of the posterior limb-bud of a four-day chick embryo. Note the presence of a head in the absence of the pelvis, and of a normal femoral curvature. Graphic reconstruction. (Murray and Huxley, 1925).

Fig. 3. Humerus developed in a chorio-allantoic graft from the basal half of the anterior limb-bud of a four-day chick embryo. Note the condyles developed in the absence of radius and ulna, the head-like structure developed in the absence of the pectoral girdle, and the hypertrophy of bone on the concave side of the abnormally bent diaphysis. (Murray, 1926.)

limbs of four frogs spinal innervation was completely excluded and the presence of sympathetic innervation, though possible, was very improbable. The only structural abnormalities were that the apophyses for muscular attachment on the ischium were always smaller than in normal animals, in some cases (but not in all) there was fusion between the femoral head and pelvis or between distal tarsals, and the limbs were somewhat shorter than normally. These experiments evidently exclude all functional factors as important agents in primary skeletal morphogenesis, but not, of course, mechanical factors caused by developmental changes.

It is not meant that no other factors than the intrinsic growth pattern of the elements themselves play any part in morphogenesis. Even in primary development, it is not known whether such concave articular surfaces as the acetabulum can develop in the absence of the associated convex element. It seems unlikely, though the experiments of Braus suggest that it may be so. Brandt (1927) made grafts of the limb anlagen of *Triton taeniatus*, mainly in the tail-bud stage. The grafted limb might push itself into the acetabulum of the host's limb, causing an enlargement and change in shape, but in other cases the graft limb was found to have formed for itself a new and additional acetabulum on the host coracoid. Similarly, it is said that in Man the displaced head of the humerus or femur, lying against the periosteum of the scapula or ilium, may cause the formation of a new acetabulum. These facts are interesting, and emphasise the lability of the skeleton during development, but do not show that the normal acetabulum has no powers of self-differentiation. The three-legged sheep described by

Jenny (1912), which had, in the limbless shoulder, a rounded knob instead of an acetabulum, proves nothing, for the knob may have been the fused proximal end of a femur the remainder of which had failed to develop, or an acetabulum may have been present in the early embryo and have been closed later by the developing bone. Hamburger's experiments show that complete absence of movement did not prevent the development of normal acetabula. Nevertheless, it is incredible that the mutual adaptation of parts, as between the two elements of a joint, could be so perfect if there were no interaction between them. Some correlation there must be, but it is not responsible for the development of gross form.

If the more recently acquired evidence is destructive of theories based on the action of function in moulding joint forms, it demonstrates with equal definition that absence of functional activity may lead to maldevelopment in later stages. In the chorio-allantoic grafts to which reference has several times been made, in the partially paralysed regenerating legs of Axolotls (Schmalhausen, 1925), in the paralysed regenerating legs of *Triton* (Brunst, 1927, 1932), and in those of Hamburger's frogs, fusion between adjacent skeletal elements was not uncommon. In the chick limbbud explants of Fell and Canti (1934) (which are described below) there was always similar fusion as a secondary phenomenon after joint formation had occurred. Brunst, too, found the fusion to be secondary, and there is little doubt that the same is true in other cases. It seems possible that fusion is especially liable to occur if the development of a joint stops short of the formation of an actual joint cavity. In the work of Fell and Canti, although well-formed articular structures appeared, they were always joined by loose connective tissue, and cartilaginous fusion subsequently occurred in all cases. In Hamburger's experiments fusion was rare, and joint cavities were formed as in normal development. It thus seems probable that functional activity, by aiding (not causing) the formation of cavities, prevents the occurrence of fusion.

In a later publication Murray and Selby (1930) described grafts of the isolated femora of six- and seven-day chick embryos, confirming and extending earlier work by Fell and Robison (1929). It was found that morphogenesis proceeded in the grafts and that the earlier conclusions were, within the scope of the later, confirmed. The grafts produced heads, trochanters, and condyles which were at least more strongly developed than when grafted. The larger size of the grafts obtained, in addition, made possible the study of certain points of anatomical detail which had not been discovered in the earlier investigation (fig. 4). Two points are of

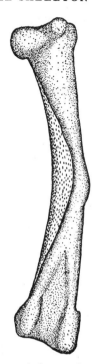

Fig. 4. A femur of nearly normal form developed as a chorio-allantoic graft after removal from the limb of a six-day chick embryo. Note the head, trochanter, and condyles, the poor development of the groove between the head and trochanter, the shallow intercondyloid fossa, the normal curvature, and the greater mass of bone on the concave side. Compare with fig. 6, p. 32. (Murray and Selby, 1930.)

interest in the present connection. However normal might be the head of a grafted femur, one feature present in the femora of normal embryonic and adult birds was always absent. This was the deep groove which runs across the top of the head of the femur and in which is lodged the acetabular ligament. The second point concerns the condyles. Even in those grafts in which the best development of condyles occurred, the inter-condyloid fossa was always very shallow on the posterior aspect of the femur. This seemed to be due to a fusion of the condyles in the mid-line posteriorly. It is evidently not possible to state with certainty the causes of these two abnormal features; at the same time it is reasonable to accept provisionally the view that the first is due to the absence of the acetabular ligament, restricting the growth of the cartilage in the region over which it lies, and the second to the absence of the tibia, but presumably, from Hamburger's experiments, less to the absence of functional movement against it than to the failure of some other relation with it.

Summarising, it may be said that the gross form of those kinds of elements which have been mentioned (mainly parts of the limb skeleton) is developed by self-differentiation, that is, under the direction of factors intrinsic in each developing element. These factors are not, however, sufficient for the production of a functional skeleton. In the early stages, when the development of gross form is proceeding, it is doubtless essential that extrinsic forces such as the growth pressure of other elements, etc. shall not deviate far from the normal conditions; it would be absurd to suggest that the intrinsic factors could produce a normal skeleton however unfavourable the extrinsic factors might be. In early stages the intrinsic factors are determinative, the extrinsic factors only important in providing conditions in which the intrinsic factors can act. In later stages, when the gross skeletal model is being refined and perfected the importance of extrinsic factors increases. It is doubtless the correlation inevitably following upon development and early functioning in close contact that causes the two components of a joint to be so perfectly adapted to one another, and evidence has been presented which indicates that various grooves, prominences, etc., of the late embryonic skeleton are probably produced in reaction to extrinsic and presumably mechanical factors.

It is now necessary to discuss the factors responsible for the development, not of the form of articular structures, but of joints at all, i.e. of the factors which ensure that the skeleton shall be a segmented structure and not a continuous branching mass of cartilage and bone. The earlier theories of joint formation generally failed to distinguish between these as two separate problems, but that such a distinction must be drawn is beautifully demonstrated in recent experiments of Fell and Canti (1934). These will now be described.

The stage of the axial condensation in the posterior limb-bud of the chick embryo is followed by one in which the future segmented condition is foreshadowed but not at once produced. There appear centres of chondrification, each representing roughly the middle of the diaphysis of a future cartilaginous element. As chondrification proceeds, the cells remote from the chondrifying centres become arranged in each element in arcs whose concavities are towards the diaphysis. In the region where a joint will later form there are therefore two opposed systems of arcs, one belonging to each element, the two systems being joined by a transition zone in which the curvature of the arcs gradually changes from that of one system to that of the other. It is across this transition zone that the division later appears; but it is noteworthy that the tissue intervening between the two articular surfaces does not disappear until about eight days, whereas the form of the surfaces themselves is established at six days. While the arcuate arrangement of cells has been proceeding, chondrification spreads from the middle of the diaphysis to the two ends, and in a 4·3 mm. bud very young cartilage matrix extends right across the line of the future joint. This interarticular cartilage matrix never develops further, and, of course, finally disappears. Its existence shows, however, that "joint formation is not caused by the presence of a layer of non-chondrogenic tissue across the site of the future articulation" (Fell and Canti).

The experiments of Fell and Canti consisted of the cultivation in tissue culture of the skeletal rudiments from the limb-buds of four-day chicks, and of the division and recombination of the rudiments in various ways. The authors, who confined their attention to the knee-joint, found firstly that the joint appeared when the whole anlage was explanted, but then discovered that if the explant contained only the presumptive knee-joint itself, without the femoral, tibial, and fibular shafts, the little structure differentiated as a single piece of cartilage without forming a joint, though some explants made an abortive attempt to do so. This suggested that the presence of shaft tissue was necessary if the epiphysial regions were to form a joint, and confirmation was obtained by explanting an entire blastema, dividing it in the tibio-fibular region and implanting between the two cut ends the isolated knee-joint region of another blastema. It was found that under these circumstances the implanted knee-joint region could form a joint provided it became, on both sides, completely incorporated with the shaft tissue of the "host" blastema. Now, as stated above, I had previously found the chick limb-bud at this stage to be a mosaic; and the next question studied by Fell and Canti was whether joint formation was a part of this mosaic. Their first result, the failure of the isolated joint region to form a joint, suggested that it was not. They next found that if the joint region was cut out of a young blastema and the cut ends of the blastema joined together again, they first healed completely and then formed a joint; but this joint lacked the specific characters of the knee-joint, the femoral condyles. Thus a joint could be formed by tissue normally some distance from the articular surface, but condyles were not formed unless the normally condyle-forming part of the mosaic was present, a result which showed that the exact position of the joint was not rigidly localised in the mosaic, while confirming my earlier work in respect of the form of the articular structures. Incidentally, I had said nothing about the process of joint formation itself. These experiments of Fell and Canti exclude as factors in joint formation not only all functional activity but also self-differentiation, and show that the presence of shaft tissue is essential.

The facts suggest that joint formation is a process enforced upon the joint-forming region by a factor extrinsic to it; but it is not yet demonstrated that there is no necessary factor intrinsic in the region and elsewhere lacking. It is true that a joint can be formed by tissue which is normally placed at some distance from the joint, but it is not known how great this distance can be, and it is probable that the ability to form a joint under the conditions of the experiment (absence of presumptive joint tissue and attachment to shaft tissue) is confined to the neighbourhood of the normal joint region, perhaps to the epiphysis. If this were the case, joint formation would come into line with the other events in the development of the skeleton: there would be an intrinsic factor localised in the mosaic, though not rigidly localised. It will be shown that there is good reason for believing the mosaic not to be completely rigid.

In 1921 and 1922 Carey elaborated a theory of joint formation based upon mechanical stresses arising in the course of development. While neither his explanation of the origin of the femoral curvature nor his theory of the development of the form of articular surfaces can be upheld, his views on the formation of the joint itself receive confirmation in the work of Fell and Canti. He regards the joint region as being first a dense cellular blastema and then as becoming a mucoid tissue; actually, Fell and Canti show that the mucoid substance is really early cartilage matrix. This tissue then liquifies. The blastema cells become flattened by the pressure exerted upon them by the growing cartilaginous diaphyses; thus the cells at the ends of the elements become arranged in arcs. This is the attitude taken by Fell and Canti; the joint is formed by the flattening of the cells in the joint-forming region under the pressure of the growing chondrogenic centres. Each such centre tends to cause the cells flattening under its influence to be arranged along arcs whose concavity is directed towards it; thus two sets of arcs appear, convex to each other, and merging by a gradual transition. The joint is formed across the zone of transition.

The next question is that of the factors determining the lengths of the different elements in the skeleton. These lengths are obviously dependent upon the positions of the joints, and on the above theory of joint formation these are fixed in part by the distances between the centres of chondrification. But if this were the only factor, every joint would be formed equidistant between the middles of the diaphyses of the two elements forming the joint. Now, in both limbs of the chick chondrification begins first in the proximal element and then in the other elements in apico-distal order. Consider the femur and tibia; the earlier commencement of chondrification in the femur should result in the zone of compression being displaced nearer to the middle of the tibia than to that of the femur. But in fact the tibia is a longer element than the femur. Thus the expectation is not fulfilled, and the order of commencement of chondrogenesis is therefore not the decisive factor in determining the lengths of the elements. As a matter of fact the difference in time of onset of chondrification in the two elements is so slight that its effect is probably negligible. Direct appeal cannot here be made to the limb mosaic because, while it is true that *in vivo* the condyles mark the distal end of the femur, the experiments of Fell and Canti show that the line of joint formation can be displaced either proximally or distally with respect to them. A joint can be formed proximal to the condyles, in which case a femur is formed from which they are lacking, or distally to them, in which case tibial material is incorporated in the femur and the condyles are abnormally large. Thus the lengths of the elements must be fixed by undiscovered factors, such as differences in the rate of extension of chondrification in the different elements, as well as by the relative positions of centres of chondrification. The latter is doubtless the principal factor in determining such gross differences as those seen when the tarsal region, for example, is compared with the thigh and shank, or with metatarsal and phalangeal regions.

It may be well at this point to summarise what is known of the limb skeleton mosaic in the chick. The earliest stage at which it has been thoroughly investigated is four days, but it undoubtedly exists at three and probably at two days. At four days the limb-bud consists of undifferentiated mesenchyme with or without an axial condensation, but without histological differentiation and with only the proximal part of the skeleton foreshadowed in the condensation if this is present at all. The mesenchyme consists of at least two groups of cells, those already determined or at least biased towards the formation of skeletal tissues (cartilage and/or bone) and those not so determined or biased. In the presumptive skeletogenous groups of cells there is already a pattern of some kind, presumably one directing its future growth, of such a degree of rigidity that, if divided into two or more parts, each part forms little more than it would do in normal development. This mosaic pattern has under its aegis all gross development of cartilaginous structure, including the form of the articular structures, but the positions of joint formation are more or less undetermined, being fixed later by factors which are themselves inherent in the mosaic: the positions and rates of growth of centres of chondrification. Thus positions of joint formation can be altered by removing parts of the mosaic, but positions of condyle formation cannot be altered in that way because the condyles are themselves parts of the mosaic. It is likely that the positions of possible joint formation may in the future prove to be limited by the mosaic, in that only certain regions may be capable of forming joints. That is, the mosaic may determine zones of potential joint formation, in which the actual position of the joint is fixed by the factors discussed above. That this is a possibility not to be excluded is shown by other evidence which indicates that the mosaic is not completely rigid, adjacent regions being separated, not by sharp lines, but by zones of transition, as suggested by me in 1926 and confirmed by Fell and Canti. Fell and Canti found that in explants consisting of the entire femoral region with a little of the tibial region attached, tibial tissue might fail to develop as such, but would become incorporated in the femur as part of the abnormally large condyles.

The preceding sections have referred in the main to the limb bones. Little or nothing is known of the factors involved in the development of any other part of the skeleton, but it is a reasonable assumption that what is true of the bones of the limbs is likely to be true also of other cartilage-replacing bones of similar general character, such as the vertebrae. In these there is reason for thinking it probable that some sort of intrinsic difference may exist, even in the stage of the earliest membranous rudiment, between the tissue destined to form the body of the cartilaginous centrum, and so to be replaced by bone, and that which will persist as inter-vertebral fibro-cartilage. Jacobson (1932 and unpublished data) finds these regions already distinguishable by the degree of aggregation of their cells and by their apparent origin from different parts of the sclerotome.

In the vertebral column it is well known that in different groups the numerical segments occupied by its various regions are not the same, and there has been much discussion as to whether there has been a shifting of function (and therefore of structure) up and down the column and from vertebra to vertebra, or whether the "same" vertebrae have retained the same function, their positions being altered as the result of the intercalation or loss of segments at various points in the column, or whether indeed the paraxial mesoderm is in different groups divided up into different numbers of segments (like a nonius) so that no one vertebra of one column can be said to be exactly homologous with any vertebra of some other column. This controversy has some interest here because if (for example) the limbs have moved about the column, it appears at first sight that one must conclude that the specialisation of a sacral vertebra depends upon the development of limbs opposite to that vertebra. In a phylogenetic sense, since the limbs are probably more immediately affected by natural selection (or other environmental factor) than the vertebrae this is probably true. When in the course of evolution the limbs begin to change their position it is probably because it is of advantage to have limbs somewhere else rather than that the limbs are moved to enable the sacrum to occupy a more favourable position. Thus, in the phylogenetic sense, the development of the sacral characters is doubtless dependent upon the position occupied by the limbs. This reasoning, however, cannot be extended to the individual development. It may be, of course, that the development of a sacral vertebra is dependent upon the presence of the limb girdle, and that any vertebra would take on the sacral characters if a girdle occupied the correct relations to it, and this is a view which is suggested by Lebedinsky (1925 a, b). The studies which this author has made of the sacrum of *Triton* seem convincingly to favour the modification theory of vertebral evolution, but the argument cannot be extended to the individual life-history until positive experimental evidence shall show that the same relationship between girdle and vertebra exists in ontogeny. It is just as probable that the same set of developmental factors which control limb development also control the modification of the vertebrae concerned and would act on the vertebrae even if the limb anlagen were removed and limb development prevented. There would then appear a self-differentiation of the sacrum in respect of the limbs and their girdles, in spite of a phyletic dependency of sacrum upon limb. Unfortunately there exists no positive evidence, though the experimental procedure does not seem unduly difficult.

Limb bones and vertebrae are bones which support the soft tissues from within; in addition to these and other bones like them there exist cartilage-replacing bones of a capsular nature, which lie outside the tissues which they support and protect. The otic capsule is the obvious example. It has been known for some years that, in certain

Amphibia, extirpation of the anlage of the otic vesicle prevents its development and that in its absence no cartilage appears representing the auditory capsule, and since this early work has aroused interest in the development of the auditory apparatus as a whole it has become clear (although authors have given but scant attention to this point) that the form of the capsule is entirely dependent upon that of the vesicle or labyrinth which it contains. Thus, if the vesicle fails to develop beyond the form of a simple sack, the capsule which forms around it has the same form (Ogawa, 1926). In the case of the optic capsule Twitty (1932) has shown that the size of the capsule is conditioned partly by the size of the eye which it encloses and partly by the amount of material available. He grafted eyes between *Amblystoma punctatum* (Axolotl) and the more rapidly growing *Amblystoma tigrinum*. The grafted eyes retained the rate of growth characteristic of the animal from which they were derived, so that one might have either a small eye in a large host or a large eye in a small host. The size of the trabecula, which forms the median wall of the orbit, was not affected by the size of the eye but there was a very marked influence on the size of the cartilaginous capsule which invests the eye itself. There was, however, also a factor depending upon the host: the amount of material available. The result was that the size of the capsule was a compromise between that typical for the host and that typical for the graft. He confirms, also, a finding by Burr (1930) of a similar effect upon the nasal capsule of *Amblystoma punctatum* in the presence of a large grafted nasal organ derived from *Amblystoma tigrinum*, but considers that in this case the possible participation in capsule-formation of cartilage-forming mesectoderm, included in the graft, has not been excluded. Burr admits this possibility in his original publication, but considers it improbable.

CHAPTER II

THE DEVELOPMENT OF THE BONY SKELETON

The primary development of replacing bones ends with the completion of the cartilage model, the secondary period begins with the first deposition of bone and the onset of erosion in the cartilage. Evidently the gross form of the bone is determined by that of the cartilaginous element which forms its scaffolding. It is therefore not necessary to discuss this matter further, but it is important that the form of the bone is not an exact copy of that of the cartilage and that under certain circumstances it may be very different. It is a general rule in the adult skeleton that, if the shape of an element is changed, as by increased curvature, or bad setting of a fracture, more bone is thereafter deposited on the concave surface. That this is true of the embryo also has been shown by me (Murray, 1926 and Murray and Selby, 1930) in chorio-allantoic grafts of entire or fragmented limb-buds of four- and five-day chick embryos and in the grafted cartilaginous femora of six-day embryos, by Landauer (1927) in the skeletons of chondro-dystrophic embryos, and by Studitsky (1934b, 1935) in grafted material similar to that used by me and Miss Selby. In the chorio-allantoic grafts the shafts of long bones were often greatly bent, presumably because of resistances encountered in the chorio-allantois to their growth in length, while similar curvatures and even fractures occurred in the long bones of chondro-dystrophic embryos, evidently caused by the failure of the pathologically soft cartilage to resist the bending stresses resulting from muscular pull. But however the curvature was caused, the effect on the develop-

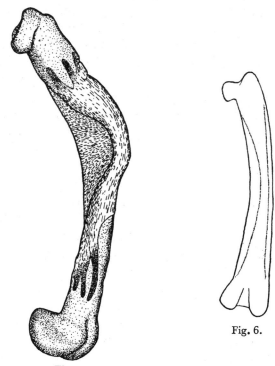

Fig. 5.

Fig. 6.

Fig. 5. A femur developed as a chorio-allantoic graft after removal from the limb of a six-day embryo. Note the mass of bone on the concave side altering the form of the element, and the condyles. (Murray and Selby, 1930.)

Fig. 6. Femur of a normal fifteen-day chick embryo. Note the groove between the head and trochanter, the deep inter-condyloid fossa, and the presence of more bone on the concave than on the convex side. (Murray and Selby, 1930.)

ment of the bony shaft was the same: there was a great hypertrophy of bone on the concave side, so that by this unequal deposition of bone, the curvature was more or less completely equalised, and the form of the element might be greatly changed (figs. 3, 4, 5, 7). The same happens in such an element as the femur, which is normally curved. Fig. 6, a diagram of the femur of a fifteen-day chick embryo, shows the presence of more bone on the concave side than on the convex, so that the curvature of the bony femur is less pronounced than that of its cartilaginous precursor. The mass of bone which fills the concavity in a greatly curved, grafted, or chondro-dystrophic element shows a very curious trabecular architecture the characteristic feature of which, as described by Studitsky, as shown in Landauer's photographs, and as frequently noticed (but never described) by me, is the radiating arrangement of the trabeculae in the hypertrophic mass on the concave side. This is clearly shown in fig. 7, a photograph from Landauer (his fig. 55). The trabeculae appear at first glance to be proceeding from a centre at the point where the element was bent, but actually it can be seen that they arise from the thin sheath of bone which immediately invests the cartilage and thence proceed obliquely towards the limiting fibrous layer of the periosteum. Such a structure as this might be interpreted by reference to mechanical factors of extrinsic origin, i.e. to the pressure which caused the bend (Landauer), or to mechanical factors brought into existence by the processes of growth and so of intrinsic origin (Studitsky), or by appeal to the necessary effects of alteration in the shape of the cartilaginous model on the normal mode of development of the bony sheath. The last two are not mutually exclusive.

Landauer holds that the architectural modifications seen

in the bent elements are produced in direct adaptation to the extrinsic pressure responsible for the existence of the curvature. He gives in his fig. 60 a trajectorial diagram of the tibia shown in his fig. 57 (as stated in the text; the reference to fig. 51 in the caption to fig. 60 is an obvious misprint), but while the arrows in his diagram correspond with the positions occupied by bony trabeculae in his fig. 57, they do not indicate the orientation of the trabeculae but are in several places perpendicular to them. Further, the bent tibia illustrated in his fig. 57 does not show the typical radial structure (being in all probability insufficiently ossified for that structure to have developed); there is no doubt, from other figures such as fig. 55 (here reproduced as fig. 7) and fig. 56, that the radial structure was at least of common occurrence. Now the most elementary mechanical considerations show that the region subjected to the most severe pressure will be that on the concave side of the bend, and that the lines of principal force will run down that side of the element (see chapter IV). Thus the kind of architecture to be expected if the structure of the bone is mechanically adaptive will be such as is indicated in fig. 9. But the direction of trabeculae in fig. 9 is entirely different from that in fig. 7; it is therefore evident that there is no direct relation between the arrangement of the radiating trabeculae and the direction of the force that bends the element.

Studitsky has recently offered a different explanation of the radiating architecture, which he studied in chorio-allantoic grafts of chick-embryo long bones. He attributes its formation to tensions set up in and below the periosteum by the growth of the cartilage. "In the inner angle of the

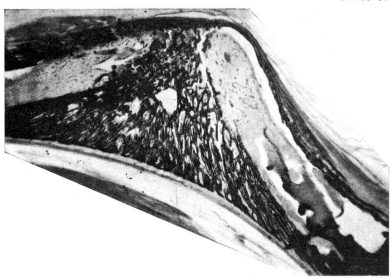

Fig. 7. The bent and fractured tibia of a chondro-dystrophic chick embryo. Note the radiating arrangement of the bony trabeculae, and compare fig. 8.

Fig. 8. Longitudinal section of the tibia of a newly hatched chicken to show the oblique orientation of the bony trabeculae. Photograph of a specimen prepared by H. B. Fell.

flexure...tensions arise in the form of two series. One of these series unites the two wings of the (cartilage) diverging with further growth; the other, secondary, is formed as a result of the withdrawing of the first series from the inner angle of the bend, and unites the fibres of this series with

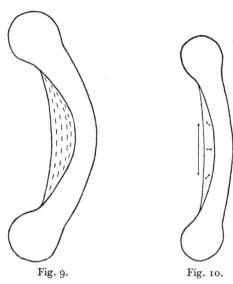

Fig. 9.   Fig. 10.

Fig. 9. Diagram of the expected arrangement of the bony trabeculae if they were oriented in the lines of pressure.

Fig. 10. Diagram illustrating Studitsky's conception of the tensions responsible for the radiating architecture.

the vertex of the angle of the flexure." In fig. 10 I offer a diagram which, if I understand him rightly, roughly indicates the distribution of tensions according to Studitsky. Such a scheme does correspond with the observed architecture. That the structure is in some way mechanically determined is strongly suggested by experiments in which Studitsky grafted cartilaginous shafts of chick long bones, denuded of perichondrium, and wrapped up in fragments of periosteum taken from the long bones of human foetuses. The human periosteum apparently formed an investment of bone around the chick cartilaginous shaft just as does the normal chick periosteum. That this bone was in fact produced by the human periosteum and not by any remains of the original chick periosteum was shown by the recognisably human chromosome complex and nuclear size. When the cartilaginous element became bent there was formed on the concave side just such a radiating structure, of human bone, as would have been formed by chick bone in the same situation. It is almost impossible to resist the conclusion that this formation, by a *fragment* of long bone periosteum, of a complete architecture harmonising with a whole element, shows the architecture to be produced by a relation between the form of the cartilage model and the periosteum around it, and it is difficult to imagine how the cartilage model could thus influence the fragmentary periost except in some such mechanical manner as Studitsky supposes. It may be remarked that Studitsky's work is very recent and as yet unconfirmed; one would wish to see a more convincing figure of the radiating architecture produced around a bent chick element by a fragment of human periosteum.

Studitsky, like Landauer, apparently regards the new architecture as produced in direct response to a new set of mechanical conditions; but examination of a longitudinal section through such a bone as a tibia taken from a late embryo or recently hatched chick shows the structure indicated in fig. 8, a photograph of a specimen prepared by H. B. Fell. As described by her in 1925 (in earlier stages), the trabeculae of bone do not run parallel to the long axis of the element but pass obliquely from the inner sheath of bone surrounding the cartilage out to the fibrous periosteum, so that they may be described as radiating from an imaginary centre in the middle of the diaphysis. In late stages these trabeculae encircle the blood vessels and become the Haversian systems or osteones, but their oblique arrangement, radiating as from a centre, can be detected not only in the late embryo but in the hatched chicken and even in the adult bird. From this it is at once obvious that the arrangement of the trabeculae on the concave sides of bent elements is a necessary consequence of a vigorous ossification following the normal lines, and not a new architecture developed in special adaptation to the forces acting. The mode of ossification which normally produces the structure shown in fig. 8 must of necessity produce that shown in fig. 7 if the element as a whole is bent and if the ossification is sufficiently active to produce the mass of bone seen. The architecture so produced is therefore the consequence of the altered shape of the element, of the large amount of bone formed, and of the pattern which the ossification normally follows; it is not produced by a new pattern of ossification manifested in response to new mechanical conditions. This is not to say that mechanical conditions play no part, but only that both the normal structure and the radiating architecture on the concave sides of bent elements are each the inevitable consequence of the same pattern of ossification and differ from one another, not in virtue of production by different mechanisms, but in being the products of identical mechanisms associated with different forms of the cartilage model. The differences in structure are the direct and immediate consequences of the differences in form of the cartilaginous precursor. It is the merit of Studitsky that he has suggested, and supported by experiment, a first approximation to a description of the mechanism responsible for the pattern of ossification which produces such apparently diverse results in different circumstances.

In Triepel's theory of cancellous architecture (1922 *a*, *b*, *c*), which will be discussed in chapter IV, the structure of cancellous bone is held to depend upon the external form of the element. Thus the interpretation just proposed for the radiating architecture in bent elements has something in common with this theory, but differs from it in not depending on a rather vague principle of "harmonische Einfügung" but on a definitely known mode of periosteal ossification. While it would be rash to extend an interpretation which is satisfactory for a particular case in embryonic development to the far more complex conditions ruling in the modification of adult bony architectures, the suggestion is obvious that more attention should be given to the effects of alteration of form upon unchanged patterns of growth. At the same time it should not be forgotten that in the embryonic dystrophic bones the architecture is not that expected on the principle of direct adaptation to mechanical forces, whereas in modifications of adult bone it frequently is.

If the architecture of the bent embryonic bones is not developed in simple adaptation to extrinsic mechanical forces, it must be considered whether the same is true, without reference to its finer structure, of the presence on the concave side of a greater mass of bone than exists on the convex side. The most obvious interpretation is, of course, to attribute this hypertrophy of bone to the heavier pressure stresses on the concave side, but here again another interpretation can be offered. In the early stages of ossification, shortly after its commencement, the cartilage model of a long bone in the chick embryo has the general form of an hour glass, the diaphysis expanding in both directions from its middle. The fibrous layer of the periosteum no longer follows the outlines of the cartilage but is seen in longitudinal sections as two almost straight lines resembling on each side the string of a bow stretched across the arc made by the concave surface of the diaphysis, fig. 11. Between it and the cartilage there is a space, which is of course widest at the middle of the diaphysis and becomes narrower towards its ends, where the fibrous sheath is again in contact with the cartilage. The osteoblasts lie scattered in this space, and here, lying between blood vessels, they build up the oblique trabeculae of bone which have already been mentioned. Clearly the existence of the space is a consequence of the shape of the cartilage, and it is also clear that the thickness of the whole sheath of investing periosteal bone is fixed by the width of the space. But in the case of an element which is greatly bent (fig. 12), the space cannot exist on the convex side, for here the shape of the element will cause the fibrous layer to lie hard against the cartilage, but on the concave side the space will be greater than usual. These facts in themselves provide a sufficient explanation of the great difference between the amount of bone formed on the two sides, rendering unnecessary any appeal to differences in the forces transmitted.

The amount of bone formed is determined by the width of the sub-periosteal space. Support for this interpretation is found in experiments (Murray and Selby, 1930) in which it was found that whereas chorio-allantoic grafts of entire

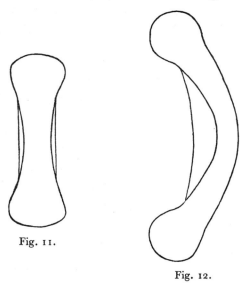

Fig. 11.

Fig. 12.

Fig. 11. Diagram to illustrate the relation between the periosteum and the cartilage in a normal long bone.

Fig. 12. Diagram to illustrate the relation between periosteum and cartilage in a much bent long bone: the sub-periosteal space is enlarged on the concave side.

isolated femora of six-day chicks developed a large quantity of bone, when the femora of seven-day chicks were grafted after division into halves (because of their great size) by transverse cuts across the middle of the diaphysis, only a very thin veil of bone was formed, although the differentiation of cartilage and its resorption in the diaphysis appeared to be in no way behind that seen in the grafts of entire six-day femora. In the half femora only one epiphysis was present, and each specimen was wide at that end and narrowed to a blunt point at the other, so that the fibrous periosteum was not lifted away from the cartilage as it is in the entire element. Hence the sub-periosteal space was only virtual and only a small amount of bone was formed.

This interpretation has thus the advantage of explaining the normal development of the thickest part of the bony sheath around the middle of the diaphysis of normal elements, the hypertrophy on the concave side of bent elements, and the under-development of bone in elements lacking one epiphysis, while avoiding the necessity of calling in extrinsic mechanical forces to account for the development of a bony mass whose fine structure must, as explained above, be attributed to other factors.

It is finally important to note that this alteration or modification of the form and structure of the bones is of general interest as an example of a development, which at first sight appears to have occurred in direct response to mechanical factors of extrinsic origin, closely resembling those associated with functional activity, but which is in reality the product of the normal pattern of growth acting on a substrate of altered form. The structure is a "growth structure" and not an "adaptive modification"; yet it is a useful structure, at least in the presence of a greater mass of bone on the side subjected to maximal pressure, and, if it occurred in an adult bone, would undoubtedly be regarded as produced in adaptation to the new mode of stressing. One is left wondering how many adaptive modifications are in reality produced, not in direct response to the conditions in respect of which they are useful, but indirectly, even coincidentally, by some such mechanism as this effect of a change of form on an unaltered pattern of growth.

The development of mammalian long bones differs from that seen in birds in the occurrence of endochondral ossification. The question therefore arises, whether it is possible to interpret the architecture of the endochondral bone in terms of the cartilaginous structure which it replaces. The problem has been investigated by Romeis (1911, and see also Mollier, 1910). As the development of the bony structure can hardly be considered in isolation from that of the cartilage, an account, based upon that offered by Romeis, is here given of both. Romeis studied the development of the tibia and also of the calcaneus of rabbit embryos. He found that in the early stages of its development the tibia is represented by a little rod of cartilage most of whose cells have their long axes transverse to that of the element itself; they are thus, presumably, disc-shaped. At sixteen days the cells at the centre of the diaphysis are becoming enlarged and show no regularity in their arrangement. Towards the two ends of the cartilage, on each side of the middle of the diaphysis, the cells are flattened and the laminae of matrix are curved, with the concavities towards the unordered region in the middle. These laminae resemble, in longitudinal sections of the element, arcs of concentric circles (fig. 13 A); they will be referred to as *b* laminae. The whole is enclosed in a perichondrium and there is no bone. By eighteen days a thin sheath of perichondral bone invests the diaphysis and part

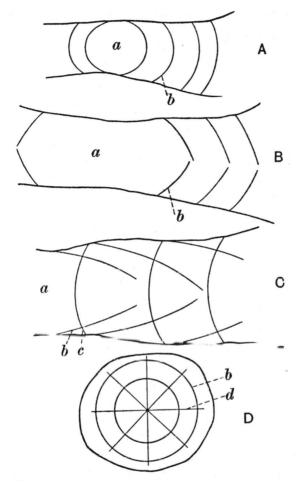

Fig. 13. Diagrams of the development of the cartilaginous architecture in the embryonic rabbit tibia. A. Sixteen days. At *a* hypertrophy of the unordered central diaphysial cartilage is beginning; the arcs *b* represent the concentric curves of cartilage cells on each side of the hypertrophic region. B. Eighteen days. The arcs *b* have assumed the form of cones. C. Twenty days. The epiphysial ends of the cones have separated from one another and the cones are becoming cylinders (in section, columns). They are now crossed by *c* laminae. D. Transverse section at the same stage, to show the *d* laminae. (Romeis, 1911.)

of the epiphysis. In the diaphysis the cells are large and vesicular. At each end of this region, the laminae which in the preceding stage appeared in sections as concentric arcs have now lost this form and resemble low, pointed arches, with the concavities deeper than before (fig. 13 B). Further, they are now united to one another by smaller cross-pieces, presumably the strengthened early matrix walls of the cells. By twenty days there has been more extensive development of the bony sheath and the middle of the diaphysis has been resorbed, so that a marrow cavity exists. The system of *b* laminae is now progressing beyond the pointed arch stage, the two arms of each arch tending to become more and more nearly parallel. The result of this is that most of the arches no longer exist as such, the two epiphysial ends of each having separated and ending separately against the unordered cartilage of the epiphysis (fig. 13 C). Considered in three dimensions the structure is now intermediate between a series of concentric cones and a series of concentric cylinders. The old unordered cartilage of the middle of the diaphysis has been resorbed and hypertrophy of the cells, soon to be followed by resorption, has commenced in the base of the region which has just passed through the stage of arches. Transverse sections made at this time (fig. 13 D) show the existence of a series of laminae, the *d* series, running radially from the centre of the element to the periphery, which are not apparent in longitudinal sections. The *b* laminae appear as concentric circles. At twenty-six days these laminae which, as stated above, were at first arcs of concentric circles, and then took the form of arches, have become parallel to one another and to the axis of the tibia, and are now the cartilage columns. They may be regarded as forming a series of concentric cylinders whose walls are not really quite parallel but slope outwards as the diaphysis is approached.

By this time the element has increased greatly in width. Now, Romeis points out that a dome which roofs over a space may be without supporting columns provided the area to be roofed does not exceed a certain maximum, but that if it does exceed this limit supporting piers must be provided. By twenty-six days or thereabouts the resorption of the diaphysial cartilage is advanced and the width of the element is increased. The unresorbed columnar cartilage forms a kind of dome roofing over the marrow cavity and, as the width of this increases with the thickness of the tibia, it becomes necessary that supporting columns should be provided. These consist at first of the unresorbed remains of the diaphysial cartilage; Romeis considers that cartilage, when not subjected to mechanical stresses, is resorbed, but that those trabeculae which carry loads from the unresorbed region to the periphery of the shaft are for a time preserved. It will be remembered that the cartilage columns are not quite parallel to the axis of the tibia but so inclined that the bases of the more peripherally situated of them run on to the shaft. The first bone is deposited upon these remaining trabeculae, and hence the architecture of this primary bone must be suited to resisting mechanical stresses and is itself determined by the architecture of preserved cartilaginous beams.

From this discussion it would appear that the cartilage of the limb elements has an architecture adapted to resist stresses transmitted from the epiphyses, at first through the axis of the diaphysis, and later on to its bony wall through the ossifying preserved trabeculae of cartilage, and that the primary architecture of the bone is directly foreshadowed by that of the pre-existing cartilage. Romeis himself, however, is careful to point out that his work leaves unanswered the question whether the cartilaginous architecture is produced by mechanical factors, by "heredity", or by a combination of the two. He favours the last view.

In reviewing the work of Romeis, Schaffer (1911) pointed out that the cartilaginous structures seen could be interpreted as a direct consequence of the growth pattern of the element, modified by the mechanical resistance offered by the bony sheath. He regards the middle of the diaphysis as a centre of centrifugal growth around which the cartilage cells arrange themselves in concentric spheres. The existence of true spheres is, however, prevented by the resistance of the perichondrium and later by the rigid sheath of bone. The spheres thus suffer deformation. Hence, proximally and distally from the middle of the diaphysis, where such resistance is not offered, the cartilage retains the form of concentric arcs (in section) until, as growth proceeds, even this becomes impossible and the arcs become pointed arches and finally nearly straight columns. In the meantime, calcification and resorption have destroyed the region of the original growth centre, and Schaffer apparently thinks of it as dividing into two and migrating physically out to the epiphyses. It is clear that such a physical migration could not occur without leaving definite traces, and no such traces exist; it is therefore simpler to suppose that the original growth centre ceases to exist and that two new ones appear one in each epiphysis. Each of these then repeats the process of centrifugal growth, producing series of concentric laminae.

There is no reason why a structure, produced in some such manner as that supposed by Schaffer, should not be adapted to fulfil mechanical functions as contended by Romeis. The two views are not necessarily opposed but may be treated as complementary. The compound hypothesis has certain advantages, for while Romeis' theory suggests that the architecture seen has a certain function without accounting for its development, Schaffer suggests a mechanism which will produce it.

The following account of the later foetal, and post-foetal development of the architecture in a long bone of Man is taken in the main from Triepel (1922 a, c).

The first spongy bone, that formed by endochondral ossification, has not the adult structure, but consists of irregularly arranged trabeculae whose long axes are parallel to that of the bone. This is to be associated either with the vascular pattern produced when the invading blood vessels, following the line of least resistance, grow along the cartilage columns from cell cavity to cell cavity, or, as Romeis suggests, with the deposition of bone along the lines of those cartilage columns which remain for a time unresorbed. The first spongiosa is destroyed before birth, by which time the cancellous bone is already of perichondral, rather than of endochondral, origin, having been formed by the excavation of the compact bone from its inner side; but the cancellous bone still does not show the adult structure. It consists of trabeculae joined by bridges, arising from the diaphysial compacta and running towards the nearest epiphysis, more or less parallel to the axis of the bone, and not crossing one another. Thus the system of "domes" and "calyces" described by Triepel (1922 a) in the adult bone (see chapter IV) is still not present, and in the human femur Triepel finds that it does not appear until one-and-a-quarter years, the diaphysis having the adult structure at three years. In the epiphysis, the architecture of the bone is for long not in harmony with that of the diaphysis. It consists at first of an irregular small-meshed network embedded in the epiphysial cartilage, but at about three years, in the human femur, the pressure tract of the medial side makes its appearance. The incongruity between epiphysis and diaphysis persists until after the union of the two; by the time this occurs the diaphysis has the adult structure and the epiphysis, after becoming united with it, undergoes a gradual transformation, finally attaining to its definitive architecture, harmoniously continuing that of the diaphysis.

This disharmony between diaphysis and epiphysis is difficult to reconcile with a strictly trajectorial theory of their development. For if the diaphysial architecture is trajectorial that of the epiphysis is not, and *vice versa*. The difficulty is perhaps not insuperable, for it may be supposed that the architecture of the epiphysial ossification, embedded in cartilage, must harmonise with the structure in which it is included. Epiphysial cartilage can hardly be regarded as a trabecular structure, but is rather one in which stresses will be transmitted in all directions; thus the included bone will be stressed to some degree from all directions, and so must have a small meshed structure not particularly strengthened in any direction. Such an interpretation may help the trajectorial theory, but a much more probable one would relate the structure of the bone to the disorderly arrangement of the cartilage which it replaces. It cannot be a coincidence that the unoriented epiphysial cartilage is replaced by an unoriented network of bone (which must subsequently change its architecture in order to become functionally efficient), and the oriented diaphysial cartilage by similarly oriented bony trabeculae, especially when it is considered that much of the development of the epiphysial ossification occurs after the onset of function. Thus it seems probable that *post hoc* may here be *propter hoc* and that the first epiphysial ossification gets its architecture either in direct dependency on the partially resorbed epiphysial cartilage or on the vascular network whose irregularity and mesh are themselves the consequence of the cartilaginous architecture which it is the business of the vessels to resorb. Subsequently, as the influence of developing function increases, the epiphysial architecture is transformed to the adult condition.

From this history there emerge points of considerable importance. Firstly, the fact that the first architecture of the bone is, in both epiphysis and diaphysis, different from that found in the adult, and that the definitive architecture only begins to show itself when the child starts to walk, suggests most strongly that the change is an adaptive one, therefore that the factor principally responsible for the architectural development of the adult spongiosa is functional activity, and that the cancellous structures produced in late foetal and early post-foetal life are determined either without reference to functional stresses or with far less dependency on them than is the case in the later modifications. An interesting experiment by Appleton (1925 a) may be cited in support of this: at an early age he removed from rabbits the gluteal muscles attached to the great trochanter. Whether the operation was performed before the onset of ossification in the trochanteric epiphysis, or shortly after it had begun, the result was the same. There was neither delay in the time of onset of ossification nor deviation from its normal progress, and there was no change in the shape of the ossified portion. The proof is perhaps not complete, but the probability is great that the architecture of the femur before walking begins is related to some other factor than mechanical stresses due to function, very possibly to the structure of the cartilage which preceded it and to the arrangement of the network of blood vessels in the marrow.

Secondly, the absence of Triepel's "dome and calyx" architecture, until it is produced during the period of adaptive modification, is unfavourable to his theory that this is the ideal structure of the spongiosa in long bones, becoming disguised by adaptation to various kinds of mechanical stressing in different elements. At the time when the effect of stress is least obvious, the ideal architecture should be clearest, but actually it does not exist, and only appears about the time when walking begins.

The changes which may be produced by the mechanical conditions of normal functioning, both in the external forms of bones and in their internal architectures, are well illustrated in the human calcaneus. What follows is derived from a study of this bone by Weidenreich (1922). Fig. 14 A shows the sectional form of the adult human calcaneus and indicates the principal features in the architecture of the spongiosa. The heel proper or Tuber calcanei rests on the ground, but the Tuberculum basale does not. The weight of the body is transmitted through the calcaneus in two directions, the chief of which is the diagonal direction from the Facies articularis posterior to the Tuber calcanei. This corresponds with the spongiosa tract (a) running in the same direction. The other direction of weight transmission is in the vertical direction from the Sulcus to the Tuberculum basale, which, although in Man it does not rest on the ground, is supported by the Ligamentum plantare longum and so carries some of the weight. There are corresponding tracts of spongiosa from the Sulcus to the Tuberculum basale (b). Other tracts are: (c) from the Sulcus to the articulation with the cuboid; (d) the ventrally convex arcs of the plantar surface, generally regarded as tension elements connecting the two main pressure tracts (but not by Jansen, 1920); (e) the tract parallel to the surface in the upper part of the heel, associated with the attachment of the tendon of Achilles; (f) the tract parallel to the surface in the lower part of the Tuber calcanei.

Now, compare this form and structure with that seen in the calcaneus of an adult whose foot, in the condition called Pes equinus, is known never to have been used for walking and never to have touched the ground (Weidenreich's Pes equinus 1). Without going into details, the neck is, in comparison with the body, high and long, while the body, instead of having a flat upper surface, rises into a convexity and from thence falls to a posterior projection whose shape is different from that of the normal. These proportions are shown by the dotted line in fig. 14 B. Internally, the spongiosa is characterised by a general weakness in its development (fig. 14 C). The system (a) is present, but the trabeculae are few in number, weak, and do not show a clear orientation. System (c) is better developed, but even in this the orientation is poorly marked. The pressure tract (b) is completely absent and the plantar system (d) as good as absent. The Achilles system (e) is absent and system (f) is replaced by an unoriented network. Thus all the systems are under-developed in accord with the disuse of the bone; almost the only recognisable systems are those which radiate from the articulation with the astragalus, for even the mere articulation produces a little pressure.

The facts so far presented indicate that both form and structure of a bone may depend upon its normal functioning. This is confirmed by a further comparison with the calcaneus of the new-born infant. At this early stage it is unfortunately impossible to say anything of the spongiosa, because the organ is still cartilaginous and contains only a small bony nucleus. The external form, however, is interesting and is indicated by the broken line in fig. 14 B, from which it is clear that in several important respects the resemblance between it and the calcaneus of the Pes equinus is closer than that between it and the normal adult condition. Contrasted with the last, the body and neck of the calcaneus are, as in the Pes equinus, higher and the neck in comparison with the body longer. The upper part of the body, instead of being parallel to the plantar surface, falls steeply from the Facies articularis superior. The adult condition is derivable from the new-born thus: the neck is retarded in growth compared with the body, i.e. it becomes relatively shorter and smaller, and the body develops especially in the diagonal line from the Facies articularis superior to the Processus plantaris of the Tuber. The whole transformation is clearly related to the conditions of stance and gait, which, because of the diagonal transmission of stress, in particular affect the heel while leaving the anterior region relatively lightly loaded. The calcaneus of the Pes equinus, never subjected to the stresses of normal life, retains a form which resembles the foetal rather than the adult state, and in its internal structure never develops the characteristic adult architecture.

From this conclusion Weidenreich goes on to a comparison between the calcanei of adult Man, human Pes equinus, human flat-foot, human new-born, and the adults of Orang, Gorilla and Baboon. The Orang is a climbing ape, therein differing from the ground-living Gorilla and Baboon. Its calcaneus has a form resembling the human new-born (fig. 14 D). According to Weidenreich the calcaneus of the new-born infant is that of a climbing animal. The Baboon and Gorilla are ground walkers, like Man, but differ from Man in walking with the Tuberculum basale of the calcaneus on the ground and the Tuber calcanei lifted above it; thus the weight is transmitted in a vertical direction from the Sulcus to the Tuberculum basale, and not, as in Man, diagonally to the Processus plantaris of the Tuber. In the human flat-foot, on the other hand, the transmission of stress resembles that in the Gorilla and Baboon, for the weight here, as in them, is carried from Sulcus to Tuberculum basale, the Tuber of the calcaneus being relatively unloaded. Fig. 14 E[1] shows

[1] In Weidenreich's fig. 13 (=my fig. 14 E) the original contains a misprint: "eines hoch-gradigen Spitzfusses" should be "eines Orangs". This is obvious from the text and from his figs. 11, 12, and 14.

the outlines of the calcanei of the adult Baboon, Gorilla, Man, and Orang, and fig. 14 F shows human flat-foot, normal adult Man, and Pes equinus. These two figures complete the story, showing on the one hand the resemblances between the Orang and the Pes equinus, and on the other that between the Gorilla, Baboon, and human flat-foot.

Thus it is clear that the outer form and inner architecture of the normal adult human calcaneus depend for their development upon the conditions of normal stance and gait, that, if these conditions are not fulfilled and the calcaneus remains comparatively unstressed, it retains throughout life a form resembling the foetal condition, and that if it is used and stressed, but in an abnormal manner, as in flat-foot, the course of its later development is changed, and changed in the same direction as is normal with such normally flat-footed animals as the Gorilla and Baboon.

The subject of the part played by functional factors in the later stages of skeletal differentiation will not be followed further here because most of our knowledge of the influence of function is derived from the study of changes produced in the already differentiated bones. These changes will be discussed, so far as the form of the elements is concerned, in the chapter which follows, while modifications of internal structure will be dealt with in the latter part of chapter IV.

Nothing has so far been said of membrane bones. In these, as will appear, the form of the bone and its internal architecture are probably developed in direct subordination to environmental factors of a mechanical nature. The

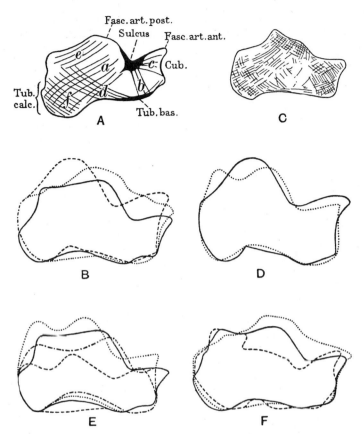

Fig. 14. A. Sectional diagram of normal human calcaneus, showing the general form and arrangement of tracts in the spongiosa. Lettering explained in text. B. Superposed outlines of calcanei of adult Man, new-born Man, case of Pes equinus, all brought to the same length; ——: adult Man; ----: new-born Man; ······: Pes equinus. C. Sectional diagram of Weidenreich's Pes equinus 1, showing the poorly developed and disorderly spongiosa. D. Superposed outlines of calcanei of new-born Man and adult Orang, brought to the same length; ——: new-born Man; ······: adult Orang. E. Superposed outlines of calcanei of adult Man, Orang, Gorilla, and Baboon, all brought to the same length; ——: adult Man; ----: adult Baboon; ─·─··─: adult Gorilla; ······: adult Orang. F. Superposed outlines of calcanei of normal adult Man, Pes equinus 1, flat-foot, all brought to same length; ——: normal Man; ----: flat-foot; ····: Pes equinus. (Weidenreich, 1922.)

following discussion will be confined to the flat bones of the skull roof with special reference to the parietal.

The development of the skull presents a group of four problems of major interest: (a) the factors determining the origin of bone in certain particular positions in the membranous roof of the young embryonic skull; (b) the differentiation of the architectural structure of the developing and adult bones; (c) the fixing of the form of the bone outlines, that is, the significance and the determination of the position of the sutures; (d) the mechanism of growth of the skull and in particular the problem presented by the increase in its radius of curvature.

Modern theoretical views of the early development of the skull roof are founded on the work of Thoma (1911, 1913), who attributed the first origin of bone, its architecture at all stages, the localisation of the nuclei of ossification, and the occurrence of the various processes by which the bone grows, all to the action of mechanical factors. Thoma compares the skull before the onset of ossification to a closed capsule which, under uniform internal pressure, increases in size in reaction to the pressure. In such a capsule the stress in the walls is everywhere and in all directions the same, assuming them to be of the same thickness throughout. Such a capsule will retain its spherical shape so long as the pressure exerted upon its inner surface is uniform, and if there is to be deviation from this uniformity part of the contents must be solid. The simplest case of a capsule under pressure which is not uniform is that illustrated in fig. 15 A, in which there are two poles at each of which the pressure is equal and greater than that exerted elsewhere. It is evident that the capsular walls are under the direct pressure of the fluid and solid contents, that this pressure is greatest at the poles, and that the walls are in addition subjected to tension in both meridional and latitudinal directions. Thoma shows that the tension stresses are greater in meridional than in latitudinal directions.

He envisages the membranous skull wall as growing in response to the internal pressure, but he supposes also that bone is caused to differentiate as a reaction to an intensification of the same factor. If this were so, and the connective tissue of the membranous skull grew in direct proportion to the pressure, there would obviously be no ossification, for the growth of the skull would always prevent the internal pressure from setting up in the wall the intensity of stress necessary for the production of bone. He therefore supposes that at a certain high intensity of stressing the growth of the membranous skull is either inhibited or fails to react in proportion, so that the actual stress developed increases. It thus reaches the critical value ("kritische Materialspannung") which causes the onset of ossification. The factor responsible for this rise in internal pressure Thoma considers to be the bending over ("Knickung") of the floor of the skull, which he therefore regards as an important if perhaps not an absolutely necessary condition of ossification in the roof.

When the stress in the membranous roof reaches the critical value ossification begins. The bone appears first in four centres, one each for the paired frontals and parietals,[1] and each centre corresponds with the position at which

---

[1] Actually there are at first two centres for each bone, but they soon unite to form a single centre for each.

the developing brain comes into contact with the roof of the enclosing skull, that is at the point of maximal stressing. Fig. 15 D shows the positions of the bony nuclei and 15 B the meridional tension stresses set up around each of these pressure poles.

Thoma's account from this time on is confined to the parietal, but is doubtless equally applicable to the frontals. The young bone has at first a net-like structure, fig. 15 C, the meshes being in the polar region isodiametric and peripherally becoming more and more pulled out in the meridional directions. The isodiametric zone corresponds to the region which is in direct contact with the growing brain, that is to the pressure pole, and is therefore subjected to predominating vertical pressure stresses while the more peripheral zone is directly affected only by the relatively slight pressure exerted by the fluid contents of the cranial cavity and is therefore subjected mainly to the radiating tension stresses. A different interpretation of this arrangement has been offered by v. Gudden (1874), who considered that the meridional orientation of the meshes in the peripheral zone was consequent upon a meridional arrangement of the vessels. Thoma will have none of this, objecting that the vessels are surrounded by so much marrow and are thus so small compared to the size of the inter-osseous meshes, that their course frequently deviates from that of the bone. Further, certain oblique trabeculae, running between the periosteal and dural surfaces of the bone, are difficult to explain by the vascular theory. Nevertheless, while v. Gudden's opinion may not be right, Thoma's reasons for rejecting it are perhaps hardly sufficient. For the mechanical determination of the architecture the

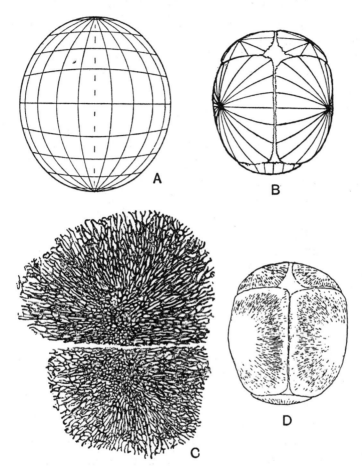

Fig. 15. A. Diagram of capsule under internal hydrostatic pressure plus pressure exerted only at the poles; meridional and latitudinal tension stresses. B. Meridional tension stresses in the bones of the roof of the human foetal skull. C. The centres of ossification of the human parietal: note that the mesh is meridionally oriented at the periphery but in the central region is isodiametric. D. The roof of the foetal skull, for comparison with B. (Thoma, 1913.)

point is probably not of primary importance, for, even if
v. Gudden is right, it is probable that the factors which
orient the vessels are those upon which Thoma relies as
directly orienting the trabeculae of bone.

As development proceeds, the beginnings of the Ebur-
neae externa and interna, the two layers of compact bone,
appear. The first to make its appearance is the Eburnea
externa, in the form of continuous lamellae of bone on the
outer side. On the inner aspect of the bone the surface is,
except peripherally, still composed of the network of trabeculae without continuous lamellae and it shows histological
signs of resorption. Towards the edges, however, deposition
of bone and the formation of superficial lamellae are to be
seen on both surfaces. Thoma interprets this to mean that
the pressure of the brain produces not only meridional and
latitudinal tension stresses and vertical pressure stresses
but also tends to cause a bending of the bone, and he
considers that this bending is of such a kind as would
subject the outer surface of the bone to still further increased
tension while compressing the inner side, so that the tension stresses are here reduced; thus the deposition of bone
is furthered on the outer surface but hindered on the inner
surface, except in the peripheral part of the bone where
the magnitude of the bending force is small. When, later,
deposition of lamellae occurs actively on both surfaces in
polar as well as in more peripheral regions, Thoma considers that the bending stresses have been reduced so that
the tensions on the inner side once more reach the level
necessary for the progress of osteogenesis. But the existence
of these bending stresses seems to be an *ad hoc* hypothesis.
There does not appear to be any independent evidence

that they exist at all, and even if the growing brain does tend to bend the developing bone, it is difficult to follow Thoma in his opinion of the nature of the resulting stresses. For surely the brain can only bend the bone if, at its area of contact with the brain, the bone is pushed outwards more than it is pushed out elsewhere, so that its curvature is altered in the sense of decreasing the external convexity, so compressing that side and extending the inner side, the opposite of Thoma's contention.

In the problem of the mode of growth of the skull, which is of minor interest for the present work, Thoma accepts the theory of interstitial growth. According to this theory the bone grows not only by apposition on the surface or surfaces and at the sutures but also by increase in size of each individual part of the bone. This concept has not received general acceptance because it is held to be incongruous with the rigid character of bony tissue. Thoma's defence of it is based upon calculations from measurements which, because of the indifferent fixation of his material and the knowledge that there were not inconsiderable differences between the form of the skulls after preservation and during life, he admits cannot be accurate.

In another section reasons have been given for rejecting theories in which the appearance of bone is regarded as a direct response to special mechanical conditions. In the particular case of the bones of the skull roof there is no direct evidence suggesting that the first ossification occurs in an indifferent connective tissue in answer to the pressures of the cranial contents, as supposed by Thoma, and it may further be asked why, if Thoma's thesis is correct, there is no ossification in the connective tissue of either the dura mater or pia mater. Weinnold (1922) has advanced other criticisms of Thoma's position. He agrees with the universal opinion that the internal pressure is responsible for the domed form of the skull but rejects the pressure-pole hypothesis as an explanation of the origin and localisation of the bone rudiments. Regarding the brain of the young foetus as a fluid-filled thin-walled bladder, he considers that the pressure which it exerts approximates to a fluid rather than to a solid pressure so that it will exert at its points of contact with the skull no pressure significantly greater than that of the surrounding fluid. More convincing than this is the evidence he adduces from cases in which the brain has failed to develop beyond a very primitive form. He cites an instance in which the brain has failed to form cerebral hemispheres, remaining a single vesicle ("Einblasenstadium"). This vesicle had a firm but, in comparison with the normal, very thin wall, and was full of fluid. Nevertheless, the roof of the skull was present and the individual bones normally arranged, both Tubera parietalia and frontalia being recognisable by the radial arrangement of their trabeculae as the ossification centres of parietal and frontal bones. Next he described two Anencephali in which "any formative action of the brain on the skull bones is naturally lacking". In spite of all this the cranial bones were recognisable, although abnormal in form and arrangement. Weinnold therefore concludes that increased stressing at pressure poles does not constitute the first stimulus to the formation of the bones of the skull roof. He considers that these first rudiments are directly inherited, mechanical factors exerting their influence only later.

Thus, while there can, of course, be no doubt that the general form of the skull roof is dependent upon its contents, it is improbable that the mechanical pressure of the growing brain can be responsible for the origin or localisation of the first centres of ossification.

Another interpretation has recently been put forward by Kokott (1933). He studied the arrangement of the principal tracts of fibrous tissue in the primordial skull of two to three months old human foetuses, and he found that these tracts are so disposed in the mainly membranous skull roof that they run between the areas later occupied by frontals, parietals, and occipitals, and are five in number. The anterior tract starts from the region of the olfactory capsule, the first paired tracts from the wings of the sphenoids ("Keilbeinflügel") and the second paired tracts from the auditory capsules and partes laterales of the occipital. These are the five highest points of the skull base, the rest of the base being at a lower level. Kokott regards the five points as of special significance as points of anchorage for the roof. From experiments with a model, which consisted of a balloon attached at five points disposed like the points of anchorage of the skull roof, and which need not be described in detail, he found evidence that the five fibrous tracts in the skull correspond in position with five systems of tension tracts in the air-filled balloon. He also observed that the form of such a balloon has, as his photograph shows, a remarkable resemblance to the form of the skull roof. He therefore concluded that the form of the skull is a consequence of the static conditions produced in it by its five-point attachment and by the evenly distributed, hydrostatic internal pressure, and that the five systems of fibrous tracts are materialisations of the principal tension lines produced by the same causes. When this stage has been reached, the primordial skull having attained its general form and developed this fibrous architecture, its wall is of course more resistant to the internal pressure along the lines of the fibrous tracts than elsewhere. In the next phase of development, for some reason which Kokott does not clearly state, the stress in the regions between the fibrous tracts reaches the value of the "kritische Materialspannung" of Thoma and ossification sets in. Kokott's reason for this seems to be an assumption that the stress in the unstrengthened areas will become greater than that in the more resistant. I am told by a competent authority that such an assumption is at least doubtful, and that the stress developed would actually be greater in the strengthened bands. It would be possible to get over this difficulty by supposing the cells themselves to be relieved of stress by the fibrous component of the tissue in the strengthened tracts—as sparrows sitting on stretched telegraph wires are not themselves under tension—but to carry at least part of the stress in the areas not so strengthened. The radiating arrangement of the bony trabeculae Kokott ascribes to the presence of radiating fibres in the ossifying regions as well as to the more direct action of Thoma's radiating tension stresses. This theory of Kokott's thus dispenses with Thoma's supposition of direct pressure exerted at points of contact with the brain and interprets the five centres of ossification as the direct consequence of the preceding fibrous architecture and the indirect consequence of the mode of anchorage of the skull roof on the base. It is an attractive idea, and its support by means of the model is impressive, at least so far as the form of the skull dome is concerned, and it is an improvement on Thoma's theory in not being contradicted by Weinnold's malformations or by his arguments against the effects of direct pressure by the brain. On the other hand it involves the unproved assumption that the factors determining the positions of ossification are wholly extrinsic, and a mechanical difficulty which seems to require a supplementary hypothesis.

The outlines of the bones are limited by the sutures, i.e. by the points at which the growing centres of ossification meet their neighbours, and it therefore appears as though the surface area and form of the bones must depend on their original points of origin, and upon the rates at which the ossification process spreads outwards from the original centres, and, of course, upon the form of the membranous skull in which they are growing. According to this view the positions of the sutures are not definite topographical boundaries but simply the places at which the bones happen to meet. This is the general opinion, but Troitsky (1932) has been led, by a series of especially interesting experiments, to a different conclusion. The experiments consisted in the removal from young rabbits, guinea-pigs and dogs of entire bones of the skull roof or of fragments of bones, the dura mater being in the majority of experiments left undamaged, and the bones, when removed entire, being carefully separated from their neighbours along the lines of the sutures. Removal of a single entire bone, e.g. the parietal, was followed by its regeneration and the sutures bounding the regenerate occupied the normal positions. Thus the gap in the skull roof was not closed by growth of the surrounding bones over the empty region but by the formation of new bone from the osteogenic layer of the dura. When the part removed contained a suture, not only did the remaining bones not grow over the gap but the regenerate had a new suture occupying the normal position. When at the removal of a bone the dura was removed with it the missing bone was not replaced and the bones surrounding the gap failed to close it, though there was a little growth at the suture edges; this in one case decreased the size of the opening from 14 × 12 mm. to 10 × 7·5 mm. in six and a half months, and in others the amount of new bone formed by growth at the sutures was of the same order of magnitude. The position of a suture is therefore more than the mere accidental meeting place of bones; it is a definite topographical boundary beyond which the growth of the individual bones does not extend. That regeneration of a missing bone proceeds from the dura is further shown by experiments in which the surface of part of the exposed dura was scraped; the bone regenerated except over the scraped area. Troitsky explains these results from the histological structure of the skull and especially of the sutures, at which the cambium (osteogenic layer) of outer periost and dura pass round the edge of each bone either without becoming continuous with that of the next, or else joining it at such a level in the vertical thickness of the skull that the junction is cut away when one or both of two bones is removed. Thus, when one bone is excised, there is no continuity between the dural cambium which remains after its removal and that of the neighbouring bones, which are left in position. The gap in the skull is therefore paved with the cambium of the dura but is surrounded by the fibrous, non-osteogenic tissue of the old suture and the cambium of the remaining bones does not cross this barrier. Troitsky undoubtedly demonstrates that the sutures are, in post-foetal stages, definite topographical boundaries, but this is not a conclusion that can with safety be extended to the first formation of sutures in foetal life. His conclusions appear to be sound once the sutures have been formed, but it cannot be held as proved that the position of sutures is predetermined before the "arrival" of the two bones at these positions. The positions of the sutures may well be fixed in the first place by the position at which the bones "happen" to meet, and each suture, once established, may thereafter be, because of its histological structure, a true topographical boundary.

Summarising, it will have become clear that not many conclusions relating to the mechanics of development in the skull can be accepted as definitely established. There is no doubt that neither the development of bone *per se*, nor the site of the ossification centres, are determined by mechanical conditions resulting from direct pressure of the brain. The characteristic radiating arrangement of the trabeculae of the early bone is indeed traceable to meridional tensions, but whether these act directly or through the blood vessels, by influencing their orientation, remains uncertain. The superficial outlines of the bones may be fixed, as generally thought, by the positions at which factors such as rate of spread of the osteogenic process cause them to meet, or the sutural lines may be predetermined non-osteogenic barriers as Troitsky seems to hold.

The problem of the mechanism by which the skull later grows in size with increase of the radii of curvature of the component bones is perhaps scarcely germane to the present study, but it should be mentioned because of the change in form which is the result of the altered curvatures. It is generally believed that each bone increases in superficial area by growth at the sutures and in thickness by deposition of bone on inner and outer surfaces. Thoma attributed the increase in radius of curvature to the combined effects of interstitial growth and unequally distributed apposition of bone on the inner and outer surfaces. Loeschke and Weinnold (1922) believe that resorption, especially on the inner surface, follows exertion of pressure by the developing brain, so that the curvature is decreased when the brain is in contact with the skull, the thinning of the bone by the resorption being compensated by deposition of bone on the outer side. Brash, in his Sir John Struthers lecture (1934), figures the medial edges of five human parietal bones of different ages, showing that resorption does occur from the inner face especially towards the anterior and posterior ends. Such unequal distribution of resorption compensated by corresponding accretion on the outer side, would evidently increase the radius of curvature. Mair (1926), unable at any stage to find evidence for extensive resorptive processes on the inner side of healthy skulls, believes that resorption of bone is practically confined to the diploe, where old bone is destroyed, while apposition of new bone occurs on both surfaces, the radii of curvature being changed by superficial inequalities in the rate of this deposition. Since there can be no doubt that extensive resorption does occur in the diploe, while this is disputed of the inner surface, it seems wise at present to give tentative adherence to the opinions of Mair. At any rate there can be no doubt that the change in curvature is brought about by unequal distribution of accretion and resorption, whether the last occurs on the inner surface, or in the diploe, or in both situations.

CHAPTER III

FUNCTIONAL CHANGES IN THE FORMS
OF BONES

The modifications induced in the form of bones by factors of a functional nature have mostly been discovered by experiments, or in the case of Man by surgical operations often of a largely experimental nature, on young individuals. Thus, while the organs studied were already more or less completely differentiated, the processes of growth were still going on, and the results of the experiments were therefore in part modifications of the last developmental stages but also frequently involved more or less drastic demolitions and reconstructions. These were necessary especially when it was required to produce a structure, like a joint, of a kind which is normally formed at an early stage of embryonic life, when both the starting-point and the biological nature of the environment were very different. Thus it should perhaps not surprise us to find, for example, that articular surfaces are brought into existence very differently when formed by the early embryo and when produced, as in the form of a pseudarthrosis, in adult life. This consideration will prevent us from extending to the early stages of normal embryonic development the conclusions reached from the study of modifications induced in later life, but permit this extension to modifications, made during late embryonic or early post-embryonic life, of the structures produced during the period of primary development.

This chapter is divided into two sections, the first dealing with the effect of functional factors on the growth in thickness and length of bones, on the development of torsions in bones, and on the limitation of growth imposed by relations with other parts, the second with changes in, and the creation of, new joints.

(1) GROWTH IN THICKNESS AND LENGTH, THE DEVELOPMENT OF TORSIONS, AND LIMITATIONS OF GROWTH

While there is little need to argue that functional factors can influence the thickness of bone, since this is universally admitted to be the case, it is of interest to mention some remarkable changes reported by surgeons who have performed the more drastic operations of bone transplantation. Thus, Bond (1913–14) removed the necrotic diaphysis of a tibia and replaced it with the transplanted shaft of a fibula. The graft attached itself to the tibial epiphysis and so changed its form (chiefly by thickening and expanding at the ends) that it practically became a tibia. Gask (1913–14) replaced three inches of the humerus of an eleven-year-old boy with a piece of fibula; the graft joined on to the remainder of the humerus and so regulated its form by thickening as to become smoothly incorporated in it, completely replacing the part removed. In some very early experiments Sedillot (1864) removed the tibial diaphysis of dogs and found that the fibulae increased greatly in thickness, so compensating for their loss.

At this point may be introduced the recent work of Wermel (1934, 1935 a–f), which although based on a single series of experiments which will be described at once, produced results requiring to be mentioned in every section of this chapter. Wermel used mainly young rabbits (at about one month) but also rats and dogs. The experiments (1934) consisted in the partial or complete removal usually of one bone of the fore-arm or fore-leg, with or without section of nerves intended to render the limb functionless, and of nerve section without any interference with the skeleton. The effects which followed were studied two to four months after the operation, and included changes in thickness and sectional form of bones, and in length, bendings and torsions, formation of grooves, and the atrophy of old and formation of new articular surfaces.

Nerve section alone was followed by disuse-atrophy but complete or partial removal of one fore-arm bone was followed by an increase in thickness in the other (1935a). The effect was of course to be referred in considerable measure to the increased stress laid upon the bone, but this seems not to have been the only factor. Thickening of the ulna occurred when operation on the radius was accompanied by nerve section, so that functional activity was considerably reduced. In a rat, too, the greater part of the tibia was removed and failed to regenerate and, although the limb remained functionless, the fibula thickened. Such results as these Wermel attributes to two factors: (a) The rearrangement of the material of the bones. This is based on the fact that all thickened bones were also shortened, and Wermel accepts the idea of an antagonism between length and thickness. (b) The removal, when one bone or a large part of it has been lost, of a mutual pressure which the two bones normally exert on one another. The remaining bone, relieved of this pressure from its fellow, is more able to realise its growth potency. At any rate it seems probable from the experiments that some factor is required to explain the results, other than the increased stress on the remaining bone, but I do not understand why Wermel insists on the action of two other factors, for his facts do not seem to require more than one, which might be either of the two he suggests.

The principal changes in the lengths of bones after total or partial extirpation of one bone of the fore-arm or fore-leg were (1934): (a) a retardation of growth in the remaining bone of the pair, shown to be a direct pathological effect and not a consequence of functional changes; (b) relative elongation of the humerus on the operated side: (c) relative shortening of the hand on the operated side. The elongation of the humerus was such that the average difference between the humeri of the operated and unoperated sides was 6·7 times the probable error and occurred in all but three of thirteen rabbits. That it was brought about by functional changes was shown by two rabbits in which, after the same operation plus the sectioning of nerves to prevent functional activity, there was practically no such asymmetrical growth of the humeri. One could wish that this combined experiment had been performed on more cases. That the absence of elongation after nerve section was not due to the failure of any trophic or similar action of innervation was shown by two rats whose operated legs lost the ability to function although the nerves were not cut; the femora on the operated side were actually shorter than those of the other side. Doubtless the functional change responsible for the elongation of the humerus was associated with the invariable shortening of the fore-arm. The relative shortening of the hand occurred in three rabbits with nerve section only, in thirteen rabbits with partial or complete extirpation of one bone of the fore-arm, the nerves not being cut, and in two rabbits with both extirpation of the radius and nerve section. Wermel gives a number of reasons why this shortening of the hand must be regarded as related to its inactivity and not to direct (trophic) effects of nerve section, effects on circulation and the like; (1) the circulatory changes were too slight, (2) there was no sign of trophic degeneration, (3) when function was weakened by operation on the bone but not eliminated by section of the nerves, the shortening was less than when activity was more thoroughly excluded. That the hands were more shortened than other parts of the skeleton is explained by the fact that in them, because of the nature of the nerve section operation, denervation was more complete, and functional activity more rigidly excluded, than elsewhere.

Other authors have denied that functional changes can increase the lengths of bones, and it is certainly true that no growth in length can occur after the disappearance of the epiphysial cartilage, but it is also true that bones which are for some reason or other not used often fail to grow to the normal length. Stieve (1927) performed, with rabbits, experiments intended to show the influence of increased stressing on the bones of the hind-legs. From one hind-foot he amputated at birth two or three toes with the metatarsals, and studied the animals eight to twelve months later. In femora, tibiae, and fibulae there was no difference between operated and control sides, but the foot bones remaining on the operated side were much thicker than on the control side, but not longer. Stieve concluded that stronger or weaker functioning does not influence the growth in length of long bones, but had to admit in view of such experiments as those of Fuld (1901) that growth in length might be affected by a sufficiently complete change in the mode, as well as in the degree, of functional activity.

Fuld (1901) removed the fore-limbs of half of two litters of puppies. The bipedal dogs had a higher tibio-femoral index (100 × length of femur/length of tibia) than the controls, thus approaching the condition in kangaroos and kangaroo-rats. In a much fuller experiment, in which the large amount of material permitted statistical treatment of the results, Colton (1929) studied the effects of bipedal gait on rats whose fore-limbs had been removed shortly after birth. It was found that the tibio-femoral index was altered, but in the opposite direction from Fuld's experiment, being lower in the operated rats than in the controls. Thus the approach was to the human and anthropoid condition and Colton makes the interesting suggestion that this may be a correlation having an adaptive significance, for the gait of the bipedal rats, like that of Man and the Anthropoids, is plantigrade, while that of kangaroos and kangaroo-rats is more digitigrade. Strictly, kangaroos are, of course, plantigrade, but since in their prodigious leaps the impetus is doubtless given by the toes they may fairly be regarded as dynamically digitigrade. Dogs, too, are digitigrade. In addition to the change in the tibio-femoral index, the ankle joint of the operated rats was wider than in the controls (due to a wider distal tibial epiphysis) and, as a result of a bend in the tibia, they were knock-kneed. Colton regarded the straddle legs, wide-spread toes and accompanying structural changes as an adaptation giving greater stability by furnishing a wider base on which to stand.

Thus, in spite of Stieve's negative result, there is no doubt that the growth in length of bones can be affected by the degree, or at least by the mode, of functioning. Lesshaft (1892) reported the case of an eleven-year-old girl who had lacked the portio sterno-costalis of the left breast muscle from birth, whose left mammary gland failed to develop, and whose left arm was shorter and weaker than the right. She did exercises for four years and the left arm became in all respects the equal of the right.

In the third of his papers (1935b) Wermel discusses alterations in the ability of bones to resist compressing and bending stresses, and the degree to which the alterations are brought about by mere increase in thickness and by changes in sectional form. Ability to resist compression and bending was not determined experimentally but was mathematically estimated from various measurements and among these the important factor of density of the bone was not included. The extent to which changes in sectional form are adaptive was determined thus: if W_a is the (mathematically determined) ability of the remaining bone of the operated side to resist compression at the end of the experiment, and W_c the resistance of the homologous bone of the unoperated side if its sectional area were equal to that of the altered bone, then $100 \dfrac{W_a - W_c}{W_c}$ represents that fraction of the difference in ability to resist compression which is due to changes in form and not merely in sectional area. If the formula represents a positive number the change in form is adaptive, if negative it is non-adaptive. With two exceptions, in all rabbits in which function persisted after loss of one bone from the fore-arm there was found to have been an adaptive change of form, the increase in resistance due only to change in *form* ranging from 3·5 per cent. to 55 per cent. of that of the normal bone. The two exceptions showed a very slight non-adaptive change. When the same experiment was performed in two other rabbits, with nerve section to prevent functioning, the change in form was in both non-adaptive. Nerve section without extirpation of bone led to decreased resistance in the radius and ulna by both decrease of sectional area and non-adaptive change of form.

That mechanical conditions can bring about the abnormal curvature of bones, especially if these are of subnormal strength, is well known in such conditions as rickets, and similar changes were described in the last chapter in chorio-allantoic grafts and in the bones of chondro-dystrophic embryos. A further illustration comes from the work of Wermel. In his experiments, a direct consequence of the partial or complete extirpation of the ulna was the assumption by the hand of a new position. The hand was rotated in the outward direction, bent over in the direction of the fifth finger, and flexed upwards, and all this so strongly as to cause a partial dislocation at the wrist. One result of the new attitude of the hand was that the rabbits walked not only on it but also directly on the distal end of the radius. The radius then became bent along the shaft of the diaphysis and between it and the distal epiphysis. There were similar changes in the ulna after removal of the radius. If, however, the hand did not become bent, or if the limb was deprived of function, the bending of the radius did not occur.

A feature of the form of bones which has excited considerable interest among anatomists is the development of torsions. In such a bone as the humerus of Man, for example, the axes of the two epiphyses are not parallel to one another, and the angle which they form is greater in the adult than at birth. The bone therefore undergoes torsion during its later development and this presents two problems: whether the torsion is produced by intrinsic factors or by such forces as muscular pull, and that of the manner of growth which brings the torsion into existence. Only the first question will be discussed here. Mention must first be made of experiments by Appleton (1922, 1925 b) in which, by operative interference or by the application of splints, the hind-legs of rabbits were rotated medially or laterally from their normal posture. The result was a torsion of the proximal end of the femur with respect to the distal end, in the direction opposite to the rotation directly inflicted in the experiments. In Wermel's work (1935 c) the new position of the hand, and the changed manner of using the limb which was its result, was followed by a twisting of the humerus of such a kind that the distal epiphysis on the operated side was turned further outwards than on the normal side. This morphological rotation was such that when the humerus was examined from the posterior aspect nearly the whole of its medial surface was visible, while none of it is seen in a normal humerus similarly placed. For this torsion the outward turning of the hand was responsible, and it was opposed by the action of muscles, which, preventing the torsion at the proximal end, were unable to prevent it distally. But the forces which they resisted they also transmitted to the shoulder blade, which in consequence also suffered deformation. Facts such as these can leave only a small doubt that normal torsions must be the result of muscular tractions and similar forces, but perhaps some possibility may remain of the existence of an undiscovered intrinsic tendency of the bones to grow with a torsion.

It is a remarkable fact, which has received no certain explanation, that although mechanical pressures of a functional kind can cause increased growth and hypertrophy of bone, other pressures can cause either atrophy or at any rate limitation of the growth of bone in the direction of the pressure. Thus a blood vessel may lie in a groove in a bony surface. Wermel (1935 d) found that when the radii of rabbits, whose ulnae had been partially or completely removed, developed abnormal curvatures, tendons which lay along them, and, because of the curvature, pressed upon them, became sunken deep into the substance of the bone, so that they lay in grooves whose edges were flanked by ridges.

The limitation which is put upon the growth of bones by the presence of neighbouring tissues is well illustrated in the human tibia. Bernhard (1924) studied the formation of the triangular cross section of this bone and found it to be due entirely to the anterior leg muscles. When these muscles contract they exert a strong pressure on the external face of the tibia and this pressure causes atrophy (or limitation of growth) of the bone and so forms the flattened surface against which the muscle lies. Even when at rest the muscles exert a lesser pressure by their tonus. At the same time, contraction of the muscles sets up tensions in the overlying Fascia cruris and, since the fascia runs into the tibial periost anteriorly, the tension is transmitted to the bone and is responsible for the formation of the crest on the anterior aspect of the tibia. Bernhard cites four clinical cases in which atrophy of the anterior leg muscles was accompanied or followed by disappearance of the crest, the tibia having then a rounded transverse section like that of a foetus. Support for this thesis is to be found in an early work by L. Fick (1857; but my information is from Weidenreich 1922, p. 466). He removed the extensor muscles from the tibiae of young dogs and found that the anterior aspect of the tibia developed, not the characteristic angle, but a rounded contour, and in experimenting on certain muscles of the jaw, he found that their removal led to a thickening of the bone with which they would normally have been in contact. In the Pes equinus 1 of Weidenreich (1922), of which the calcaneus has been discussed above, the extensor muscles were absent from the lower part of the shank, having been converted into a fatty and useless mass; correspondingly, the anterior crest of the tibia was represented only by a small ridge in the middle and distal thirds, while the proximal third was evenly rounded. The abnormal tibia thus resembled that of foetuses and of children in their first year, in which too the cross section is rounded, the adult form only appearing when walking has begun. In the adult Orang the form of the cross section is more rounded than in Man, but in the Gorilla, which walks more and climbs less than the Orang, so that its tibia is differently used, the crest is more developed. The fact that the Orang has the least developed tibial crest among the Anthropoids, coupled with its climbing habits, indicates that not only the presence of muscles but also the manner of their use is of significance in determining the development of its form.

It should be stated that a different interpretation of tibial form was put forward by Hirsch (1895) in a publication which I have not been able to see. It is, however, summarised by Biedermann in Winterstein's *Handbuch der vergleichenden Physiologie*, **3**, 1, 1. Hirsch investigated the forces acting on the tibia in various positions of the body, and found that in standing on one leg with the knee extended the tibia tends to be bent with the concavity forward, while with the knee flexed the forces acting tend to bend it with the concavity backward, and in standing on two extended legs the bend is in the outward direction. The characteristic triangular cross section of the tibia Hirsch regards as adapted to resisting bending in these planes. The clinical cases cited above are explicable from either point of view, for since the muscles were atrophied the legs were presumably not functional and the bending stresses therefore not acting. It should however be stated that Benninghoff (1935) regards the triangular cross section as not the best possible form from the point of view of the tibia, for the performance of whose functions a round section would, in his opinion, be preferable. In agreement with Bernhard, he regards its form as impressed upon it by its muscular environment. From this point of view, the tibia is an excellent example of a structure developed in compromise between the mechanical demands on the skeleton and the spatial needs of the muscles.

(2) CHANGES IN JOINTS AND THE CREATION OF NEW JOINTS

To study the effect of movement on the form of the femoral head, R. Fick (1921) united the two opposite legs of dogs and rabbits, as by passing wire through them. Thus, whereas normally the hip joint allows movement in all directions, after the operation movement was limited to a single antero-posterior plane. The principal result was a change in the form of the femoral head from spherical to cylindrical.

In an operation by Wachter reported by R. Fick (1921), the muscular attachments in the shoulder region of a rabbit were altered, and the joint opened and "freshened". A new joint was formed, but the scapula bore the head, the humerus the pan into which it fitted; the result accorded very well with Fick's theory according to which the form of an articular surface is closely associated with the position of the attachments of muscles working the joint.

Wermel's experiments (1935c) involved a number of drastic changes in articular structures. As already stated, the hands of the rabbits which had lost part or the whole of the ulna by operative interference took a new and abnormal attitude and were partially dislocated at the wrist. The bending over of the hand in the direction of the fifth finger pulled the carpus away from its articulation with the radius but forced it up against the outer side of the ulna. Normally (fig. 16 A) the radius has a broad articular surface with the naviculare and lunatum, and the ulna a small surface on the little capitulum ulnare articulating with the triquetrum (ulnare or cuneiform). In the modified joint (rabbit 2, fig. 16 B) the radius, which, owing to the partial dislocation now hardly touches the carpus, has its joint surface reduced to a rounded prominence (2) like a normal capitulum ulnare, while the real capitulum ulnare (3) is fused with a new mass of bone

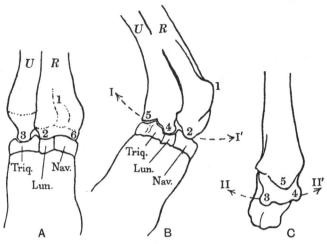

Fig. 16. A. Diagram of normal carpal region of rabbit. B. Lateral view of carpal region of Wermel's rabbit 2. C. Ulna of rabbit 2 from behind. R=radius, U=ulna, Triq.=triquetrum, Lun.=lunare, Nav.=naviculare. The numerals indicate corresponding positions in the three diagrams. I–I', II–II' explained in text. (Wermel, 1935c.)

(fig. 16 B and C, 4), forming with it a large roll-like joint (fig. 16 C, 3 + 4). In addition, another new prominence (fig. 16 B, 5) overhangs the triquetrum. Thus the dislocation brought about two sets of changes: disuse-atrophy of the greater part of the articular surface of the radius and, on the ulnar side of the joint, a depression between 4 and 5 (fig. 16 B), formed by the pressure of the triquetrum, while deposition of new bone has filled up empty spaces and formed these two prominences. The result is an immovable joint, because in the direction I—I' movement is prevented by the prominences 2 and 5, while at right angles to this, II and II', it is prevented by the roll joint 3 + 4 (fig. 16 C). This immobilisation is an adaptive change to the extent that it increases the efficiency of the hand in the new circumstances by preventing it from being bent under when subjected to the pressure of the body weight.

Very considerable alterations occurred at the elbow-joint, in consequence of the downward slipping of the proximal fragment of the ulna, which was pulled by the outward turning of the hand and fore-arm and was fused with the radius. The result was a partial dislocation of the humerus (fig. 17 B), leaving between it and the olecranon an empty space (fig. 17 B, a), while the articular surface for its reception under the olecranon (fig. 17 A, d) disappeared altogether. Instead, two new joint surfaces appeared on the olecranon (figs. 17 B, c; 17 C, b and c) and corresponding with these were two new surfaces on the humerus (fig. 17 D, e and f), lateral to the fossa supra-trochlearis. It is evident in all these changes, as also in those at the wrist, that depressions and new articular surfaces were formed at points where bones rubbed against one another, while joint surfaces which no longer worked upon their fellows suffered atrophy.

In his sixth paper (1935 e) Wermel describes the regeneration phenomena which he observed in the course of his experiments. The most interesting point is what appears to be a very extraordinary new observation: the regeneration of articular surfaces independently of the bones upon

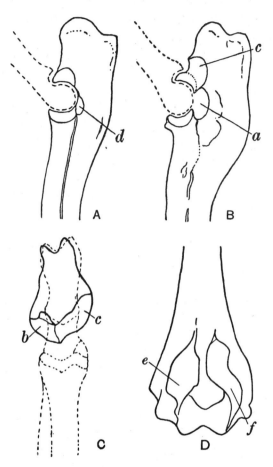

Fig. 17. A. Diagram of elbow-joint of normal rabbit. B. Diagram of elbow-joint of Wermel's rabbit 2, from medial side. C. Upper part of ulna of rabbit 2 (thick lines), superposed over normal ulna (dotted). D. Lower end of humerus of rabbit 2, from behind. *a*=space between ulna and humerus; *b*, *c*=new articular surface for humerus on olecranon; *d*=normal articular surface for humerus; *e*, *f*=new articular surfaces for ulna on humerus. (Wermel, 1935*e*.)

which they would normally be borne. Wermel takes, as an example of events in all cases of removal of the radius, rabbit 13. In this the radius did not regenerate but in the region of the elbow-joint there formed a large accumulation of cartilaginous tissue ("knorpelartigen Gewebes") which assumed the form of a normal articular surface of a radius and formed a support for the humerus.

The regeneration of joints leads immediately to the subject of pseudarthroses. These are joint-like structures which appear in places where normally there should be no joint. Most commonly their formation follows the failure of the two ends of a fractured shaft to reunite, but they may be formed by the appearance of a new division across the callus which has already joined the fractured ends, or across a piece of bone transplanted in the place of a previously existing pseudarthrosis in the hope that it will effect a permanent union of the two halves, or, perhaps most remarkable of all, by the spontaneous division of the shaft of the apparently healthy radius when a pseudarthrosis already exists in the ulna. But the mere failure of a fracture to heal, or the mere appearance of a new fracture in callus, graft, or ulna, does not constitute a true pseudarthrosis. This is the result of a further series of changes which may lead to the formation of a structure so exactly mimicking a normal joint that the first half of the word "pseudarthrosis" does it less than justice. In the most perfectly formed pseudarthroses, the apposed ends become expanded like the ends of a normal bone, one end forms a head-like and the other a pan-like articular surface, these surfaces become covered with cartilage, and the whole structure is enclosed in a capsule, with ligaments and a synovial fluid. One of the most interesting of known pseudarthroses was described by Maier (1930) (fig. 18). It followed a war wound which shattered the humerus at

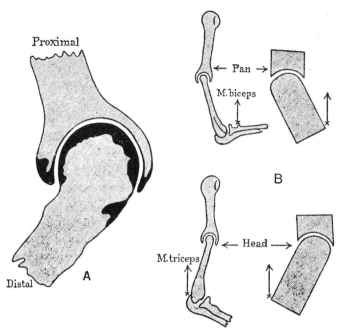

Fig. 18. A. Diagram of Maier's pseudarthrosis. Proximal end above, cartilage black. B. Diagram to illustrate the relation between the pseud-arthrosis, and the tendon attachments of M. biceps and triceps, and the effects of point of actions of forces on gypsum blocks rubbed against one another. (Maier, 1930.)

about the middle of its length. Fifteen years later the pseudarthrosis was a very well-developed spherical joint, with a rounded head on the distal segment and on the proximal segment a spherical concavity which almost completely enclosed the head. This, incidentally, is a form of joint not normally found in Man. The head was covered with cartilage but the concave component had cartilage only around the lip, as indicated in the diagram. There was a fibrous capsule which contained about 1 c.c. of fluid. The false joint was very mobile in all directions and the patient had to abandon his occupation of blacksmith, but during the fifteen years had performed other kinds of manual work in which he learned, by various tricks, to make use of the injured arm. The form of the pseudarthrosis accorded well with R. Fick's views of the relation between joint form and position of muscle attachments, for the biceps and triceps, the important muscles in moving the "joint", are attached remote from it, the proximal segment of the humerus being regarded as the stationary component.

Wermel (1935 e), in the course of the experiments already mentioned, found and briefly described an interesting pseudarthrosis in a dog's tibia. A large section had been cut out of the tibia, leaving the epiphyses intact. In the course of five months the ends of the two tibial fragments formed a movable pseudarthrosis, the upper part being convex, the lower concave. The remarkable point was that the distal end of the upper segment of the tibia was separated from the remainder of the segment by a plate of cartilage-like tissue and the structure was externally indistinguishable from an incompletely ossified epiphysis.

The formation of a pseudarthrosis involves two processes. The first is the failure of an existing fracture to unite or the appearance of a new division in a previously undivided bone, while the second is the elaboration of the joint-like form and of accessory structures such as the fibrous capsule. The factors concerned in the first process have been studied especially by German surgeons (Bier, 1923, Willich, 1924; Lexer, 1922; Martin, 1933–34; Müller, 1921, 1922; and others). Unfortunately the authors are in disagreement, and it seems probable that a number of factors act, different ones being decisive in different cases. Bier (1923) showed that such gross mechanical action as exists when the two ends of a fracture are allowed to move upon one another are not necessary to pseudarthrosis formation, for this can occur in plaster casts. Indeed, spontaneous division of a callus, or of a bone graft, or sympathetic division of the radius may occur under the same conditions of apparent exclusion of mechanical forces. It would, however, be rash to conclude from this that all mechanical factors are excluded by plaster casts, or that they are ever quite without effect in pseudarthrosis formation, for a cast does not necessarily prevent all movement, however slight, nor does it prevent isometric or tonic muscular contraction. An experiment by Müller (1922) indicates very strongly that such forces may be the decisive factor. He grafted a piece of bone across the ankle-joint of rabbits, thus immobilising the joint; later the joints regained their mobility as a result of pseudarthrosis formation in the grafts, exactly at the position required. Apart from mechanical factors in the sense of actual movement or muscular pull, the interposition of soft tissues between the ends of a fracture may lead to pseudarthrosis formation, but, if the callus is able to penetrate the barrier, may not. Lexer (1922) considers the vascular supply of particular importance, as determining whether callus shall be formed, and how much, pseudarthrosis occurring more readily if the callus be weak than if it be strong. Willich (1924) attaches more significance to the relative parts played, in callus formation, by the periosteum and endosteum, pseudarthrosis being more probable if either of these fails in its role, less so if both are active. Mechanical factors, such as those acting when a fracture is not completely immobilised, may assist in pseudarthrosis formation, especially if the callus is weak, but may not prevent good healing if it is strong.

Turning to the second process it is of interest that the form of the articular surface develops gradually; the pseudarthrosis may appear first as a narrow transverse area of resorption, then this becomes a split, while the elaboration of the concavo-convex articular surfaces follows later. Thus it seems probable that, whatever factors are responsible for the original cleavage or failure to unite, these factors are probably not responsible for the modelling of the articular surfaces; otherwise, one would expect the cleavage planes to be joint-like from the first.

It is quite impossible to resist the conclusion that the factors which lead to the formation of the joint-like structures must be mechanical, for a considerable quantity of evidence shows that, if the severed end of a bone is not in contiguity with another end, it does not expand and become joint-like. Thus, the proximal stumps of bones whose distal ends have been amputated do not become like joint surfaces. The same conclusion applies to the formation of new articular surfaces in Wermel's experiments, for here there can be no doubt that they appeared just at those points at which bony surfaces rubbed against one another.

Thus, while the normal articulations are brought into existence during the period of primary development by factors quite other than the functional, their production in later stages is invariably to be attributed to causes of a mechanical and functional nature.

Wermel, in his fourth paper (1935 c), gives an interesting table, most of which is here reproduced, summarising the complicated series of changes brought about in the limb as a result of the changed manner of its use, itself the consequence of one single primary factor, the altered position of the hand. The direct result of the partial dislocation

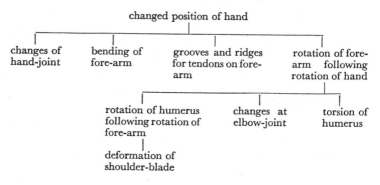

of the hand is the series of changes at the wrist; of the habit of walking directly on the distal end of the radius, the curvature of that bone; while the curvature itself leads to the formation of the grooves in which the tendons lie. The rotation of the hand causes corresponding rotation in the fore-arm and humerus, and so the torsion of the humerus, the deformation of the shoulder blade, and the partial dislocation of the elbow with the resultant changes in it. The thickening of the surviving bone of the two in the fore-arm, the relative elongation of the humerus and the shortening of the hand fall into different chains of causation, each expressing increased or decreased intensity of stressing. The series of changes included in the table forms a remarkable group of correlations, but is paralleled by the alterations obtained by Fuld (1901) and Colton (1929) in their experiments with two-legged dogs and rats, where too all parts of the limb were affected, though less drastically, again by a change in the manner of gait. Evidently the fact cannot be over-emphasised that the skeleton, and with it the muscles, tendons, and ligaments, forms a functional unit, alteration to any part of which may, by bringing about changes in the use of the part, affect regions far beyond that directly concerned in the alteration and perhaps even throughout the entire skeleto-muscular apparatus. The profound difference between this and the self-differentiating skeleton of the early embryo is so striking as scarcely to need mention.

At the beginning of his work (1934) Wermel lists four processes or mechanisms which he regards as at work in the production of the modifications he describes. These are: (1) hypertrophy due to activity (e.g. thickening of radius when doing the work of both radius and ulna); (2) hypertrophy under conditions of inactivity ("Inactivitätshypertrophie", e.g. thickening of the radius or ulna after extirpation of one of them with loss of function); (3) atrophy due to activity (e.g. the atrophy of bone under tendons, leading to groove formation); (4) atrophy due to inactivity (e.g. of long bones after loss of function or of disused articular surfaces). The first and fourth no one will doubt, and the third will be admitted except that it is doubtful whether the rubbing of a tendon on the surface of a bone can fairly be compared, as the use of the term "Activitätsatrophie" seems to suggest, with functional activity. The second seems to me less well established, for it is difficult to be sure how far functional activity was excluded. Wermel recognises that his nerve sections did not completely stop all functioning and even in the rat with the operated non-functional limb it is perhaps permissible to doubt whether it never used the limb even for passive support. Nevertheless Wermel's facts do make it probable that thickening of the bones could be brought about by some factor other than increased activity. I would prefer a rather different list: (1) Acceleration of growth in thickness and less commonly in length, in response to increased functional stressing. (2) Disuse atrophy. (3) Atrophy, or at least limitation of growth, imposed upon the skeleton by pressures of a kind different from those which, having regard only to the skeleton, can be regarded as functional (e.g. pressure from tendons, the effect of muscular pressure on the form of the tibia). It is interesting to compare this with Nauck's (1926) conception of "Ausdehnungshemmung" in the early embryo. Wermel's "Inactivitätshypertrophie" might, I think, be regarded provisionally as a special case of growth after removal of a normally present limitation. There must also be another process (4) at work in producing the modifications: the new creation of tissues. The formation of new cartilage-covered articular surfaces, as in Wermel's experiments and in pseudarthroses, especially when associated with new joint capsules, ligaments, etc., cannot be classified as mere hypertrophy due to increased activity, for a hypertrophy is the increase of something present, while the formation of a new joint requires excavation of a depression (atrophy), elevation of prominences (hypertrophy), and creation of the cartilaginous surface.

Finally, all these factors must be operative in the later stages of normal development, for at that time the skeleton consists almost entirely of the material basis, bone, on which the factors we have discussed are shown to operate, and simultaneously the intensity of functional activity, and the mechanical forces by which it is accompanied, are playing an ever greater and greater part in the life of the organism. The fourth factor indeed, the creation of new tissues in response to functional needs, cannot commonly be of normal occurrence, but even this is possibly represented in the development of cartilaginous articular surfaces on membrane bones and probably by the development of accessory structures around all kinds of joints. Thus the statement made in the preceding chapter, that functional and mechanical factors play an increasingly active part in the later stages of normal development, receives support from the facts presented in the present chapter.

CHAPTER IV

THE MECHANICAL STRUCTURE OF BONES

(1) The Structure of Normal Bones

The greater part of the work which has been done in interpreting the form and structure of bones in relation to their function has been concerned with the long bones of mammals, especially Man, and particularly with those of the limbs. For this reason it is to these that the following discussion will in the main relate.

A long bone from the limb of a terrestrial mammal has the form of a hollow cylinder, whose wall is formed of compact bone, which is thicker about the middle of the bone than towards the ends, where, indeed, it may nearly disappear. The ends of the bone are expanded, and it is in the ends especially that the cavity contains the spongy or cancellous bone. This generally shows a definite tendency towards a certain arrangement or architecture, the trabeculae being in some directions much stronger and longer than in others.

It is these facts, but particularly the details of the cancellous architecture, that various theories have sought to explain.

A. *The trajectorial theory*

When a block of material is subjected to forces it temporarily changes its shape because of its elastic properties. These small changes in shape are called "strains" and the internal forces which resist the deforming action of the external forces are called "stresses".

Consider a simple block (fig. 19 A) supporting a load. The block will be shortened vertically, but it will not be seriously reduced in volume, so it expands horizontally (fig. 19 B); i.e. a small circle imagined on any vertical plane in the material becomes an ellipse with its major axis horizontal and its minor axis vertical (fig. 19 C). Actually, solid bodies have three dimensions, and we ought therefore to consider a sphere deformed into an

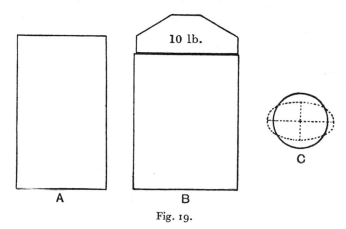

Fig. 19.

ellipsoid, but in what follows we will for simplicity consider only two dimensions. The argument and general results are not altered.

In the block, a simple compressive force in one direction produces strains in all directions, but there are at any point two directions at right angles along which the strains are respectively a maximum extension and a maximum compression (fig. 19 C). These directions are called the principal axes of the strains. Now a strain is a linear movement, i.e. a matter of direction, but a stress is a force measured across an area of a plane. The stress across any plane in a body is a simple linear function of the strain in all directions and the plane of principal stress at any point is at right angles to the principal axes of the strains. That is, the axes of principal stresses at a point are identical with the principal axes of the strains.

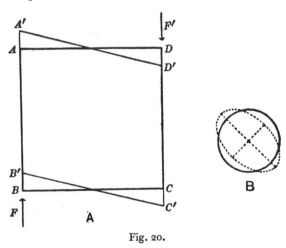

Fig. 20.

Now consider a square block *ABCD* (fig. 20), which has a force *F* acting upwards along the side *AB*, and a balancing force *F'* acting downwards along the side *DC*. The square will tend to deflect into a shape such as *A'B'C'D'*. This is known as a shearing strain. The diagonal *DB* shortens to *D'B'*, and the diagonal *AC* lengthens to *A'C'*. Thus a small circle becomes an ellipse (fig. 20 B) with its major axis tilted down from left to right. However, we see there are still two principal axes of strain, which are at right angles and are directions of maximal lengthening and shortening.

We can now state the general results of the elastic properties of material as follows. When a block is subjected to a system of forces, i.e. tensions, pressures, and shearing forces, however complicated, then at every point in any plane section of the material:

(a) There are two directions of principal stress.

(b) These directions are at right angles to each other.

(c) The stress in one direction is a maximum tension (or minimum pressure) and in the other direction is a maximum pressure (or minimum tension). Generally in all the cases with which we have to deal we can say that the two directions of principal stress correspond to the directions of maximum tension and pressure.

(d) In the directions of principal stress there are no shearing stresses; but in any other direction there will be shearing stresses.

It is thus possible to draw lines on any imaginary section of a body along the directions of principal stress. These lines may be straight or curved, but at all points in the body the two sets of lines remain at right angles to each other, and the closer they are together in any region, the greater the principal stresses in that region, and vice versa. Following the usage of German authors, these lines are known in anatomical theory as "trajectories".

Consider now a beam *ABCD* (fig. 21) fixed in a wall by the end *AB* and loaded uniformly along its whole length or not loaded at all (except by its own weight). The beam will bend, so that the upper side becomes a little longer, the lower side a little shorter. The strain in the upper half of the beam is one of extension, in the lower half one of compression; it is greatest in the superficial layers on the two sides, and midway between the upper and lower surfaces of the beam there is a neutral layer which is neither extended nor compressed. It is an experimental fact that in elastic bodies under any kind of stress the stress is proportional to the strain, and therefore the tension and pressure stresses are, like the strains, greatest at the upper and lower surfaces respectively, and decrease towards the

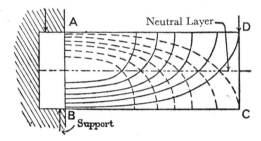

Fig. 21. Trajectorial diagram of beam attached at *AB* and uniformly loaded. (Koch, J. C., 1917.)

neutral layer. If on any vertical section of the beam we draw the lines of principal stress, the pressure lines (trajectories) must start on the side of application of pressure (*AD*) but run mainly along the side on which the beam suffers compression (*BC*). Thus they will arise from points on *AD*, at right angles to it, will describe a (parabolic) curve crossing the neutral layer at 45 degrees, and will run along near the surface (*BC*) till they meet *AB*, again at right angles to it. The tension trajectories will arise on the side opposite that on which the pressure is exerted, but will run along the side which is extended (*AD*), crossing the

neutral layer at 45 degrees and crossing the pressure
trajectories at right angles. If now it be supposed that the
beam is supported at both ends and evenly loaded throughout its length, as in fig. 22, *AD* becomes the side of compression, *BC* that of extension, and the trajectories run in
the manner shown.

Evidently in building a structure to resist the deforming
action of an external force, it is advantageous to arrange
the material so that its greatest mass is concentrated where
the stress is greatest and that it is oriented along the lines

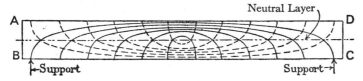

Fig. 22. Trajectorial diagram of a beam supported at both
ends and uniformly loaded. (Koch, J. C., 1917.)

of principal stress. Since there are no shearing forces in
these directions, while in other directions there are shearing
forces, the disruptive effect of shear is evaded. A cellular
or lattice structure is therefore possible. If the parts of a
lattice or cellular structure did not occupy the lines of
principal stress they would be subjected to shear, and would
therefore require to be more heavily built. Thus by putting
the material in the right places the structure is given the
maximum strength obtainable with the amount of
material used, or alternatively, is given the required
strength with the minimum of material.

The fundamental idea in the trajectorial theory of bone
structure is that the trabeculae of cancellous bone follow

the lines of trajectories in a homogeneous body of the same form as the bone and stressed in the same way. They are thus regarded as actually materialised trajectories. This was the classical theory advanced by Culmann and Meyer (Meyer, 1867), adopted by Wolff (1870, 1892, 1899), Thoma, and Roux (1885), and supported to-day, among others, by Petersen (1930). Historically, it takes origin from the observation of Culmann that the preparations and drawings by Meyer of the cancellous architecture seen in a frontal section of the proximal end of the human femur closely resembled the trajectories which an engineer would draw in a crane of the same general external form as the femur (Fairbairn's crane) and of homogeneous structure.

Fig. 23 illustrates the architecture of the humerus and radius of the dolphin. It is probably shown better in the radius (fig. 23 B and C) than in the humerus (fig. 23 A), but unfortunately the only figure of the radius known to me is Roux's (1893) rather unsatisfying diagram reproduced (with Benninghoff's additions, 1927) in fig. 23 B and C. Both humerus and radius have three characteristics: (a) Practically the whole bone is full of spongiosa; but actually there is a canalis nutritius along the axis, following the neutral layer. This absence of a marrow cavity is common in marine animals. (b) The compact bone is thickened on the dorsal and ventral aspects but, as fig. 23 C shows, not anteriorly or posteriorly; Schmidt (1899) states that the same is true of the humerus. (c) The architecture of the spongiosa resembles the trajectorial diagram of the bent beam (fig. 22); both radius and ulna and the shaft of the humerus show two systems of trabe-

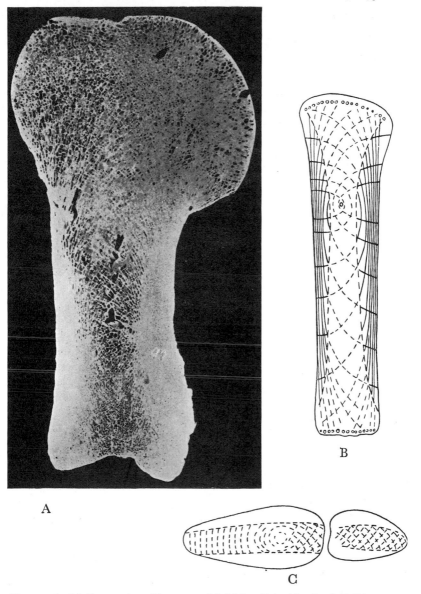

Fig. 23. A. Median section of humerus of dolphin. (Schmidt, 1899.) B. Diagram of the spongiosa structure of the dolphin radius, after Roux, with the addition of the "split lines" in the compacta. (Benninghoff, 1927.) C. Transverse section of radius and ulna of dolphin, from Roux. Note the dorsal and ventral thickening of the compacta.

culae, each of which starts and ends on the same side, as do the two sets of trajectories in the bent beam.

The dolphin is a marine animal which never comes on land; thus the limbs are relieved of the pressure of the body weight, which, in terrestrial animals, they must support. Further, they are not the main organs of propulsion, a function performed by the expanded tail, but are used as horizontal rudders, like those of a submarine, for controlling movements in the vertical plane. When the undulating movements of the dolphin, or of the closely related porpoises, are considered, it is evident that the limbs are subjected to bending stresses in a roughly dorso-ventral direction. The stream-lined form relieves them of any considerable pressure, in an antero-posterior direction, exerted by the water. In the thickening of the compacta on the dorsal and ventral sides, and in the close agreement between the arrangement of the spongiosa trabeculae and the trajectories of the bent beam (fig. 22) these bones are a beautiful example of a mechanically suitable architecture.

In terrestrial animals the limbs have the two functions of supporting the body and of moving it from place to place. The human femur is the bone which has been most thoroughly studied, especially by Culmann and Meyer (Meyer, 1867) and by Wolff (1870, 1899), while the most thorough mechanical and mathematical analysis to which it has been subjected was that made by the American anatomist J. C. Koch (1917). It is not without interest that, as early as 1838, the English anatomist Ward recognised the resemblance to a crane (actually, he referred to a "bracket") but, according to Wolff (1892), chose the wrong crane because the correct analogy, the Fairbairn crane, had not yet been invented. The femur (fig. 24) is a long bone, rounded in transverse section, hollow and expanded at the ends. The spongiosa is confined to the two ends of the bone, and in these regions the compacta is very much thinner than in the middle of the diaphysis. Now, a flanged girder (fig. 25 A) is a structure adapted to resist bending in one plane only, while a girder of square cross section (fig. 25 B) resists bending in two planes. The flange which connects the pressure and tension components of the first is represented in the second by the two walls which are at any time tangential to the force acting. Evidently, if the girder is to resist bending in all directions it must be of circular section, like the femur, which must in fact offer such resistance owing to the constantly changing position of the centre of gravity of the head and trunk.

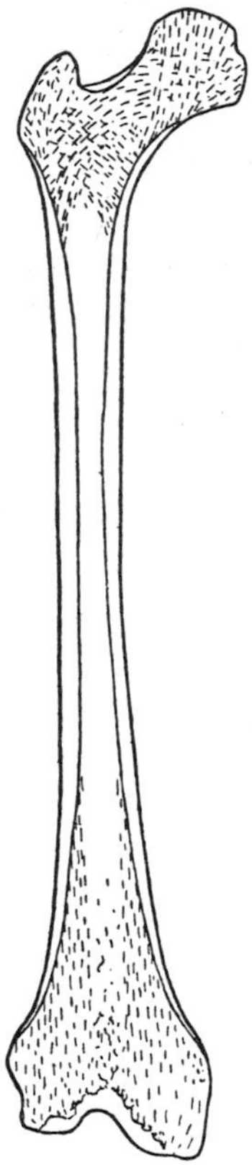

Fig. 24. Diagram of a frontal section of the human femur, in part after a photograph by J. C. Koch (1917).

The terminal expansion of the femur (perhaps better exemplified in the tibia) is related to the greater stability obtained when the ends of two superposed columns are relatively wide.

The thickening of the compacta is frequently regarded as representing a fusion of the spongiosa trabeculae, but Gebhardt (1910) pointed out that this could not be the whole explanation because the maximal thickness of the compacta is reached at the middle of the diaphysis, remote from the point of entrance of the last trabeculae.

The thickening is a response to the existence, at the middle of the diaphysis, of a region of danger, or of least security. In a bent beam the stress in the beam is greatest where the trajectories run most closely together. Thus in fig. 21 the beam, if excessively loaded,

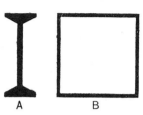

A   B

Fig. 25. Explanation in text.

will break along the line *AB*, and in fig. 22 across the middle. If one breaks a straight walking stick by pressing it against the ground or by bending it from the ends, it snaps at about the middle of its length. Similarly the femur, were the compacta not thickened around the middle of the diaphysis, would have there a point of danger, where it would break very readily. The diaphysis of the femur may thus be regarded as a hollow column, with its ends expanded for greater stability and with its walls thickened in the middle of its length to increase its resistance to bending. In the epiphysial region the compacta becomes so thin as nearly to disappear, the bone consisting practically entirely of spongiosa.

Benninghoff (1925a, 1927, 1930) has used an ingenious method for discovering the general course taken by fibre bundles in the general lamellae of the superficial layers of the compacta and of the arrangement of osteones in the deeper layers. Pricks are made with an awl and a stain is then rubbed in. On removal of the stain from the surface of the bone it remains in the small splits made by the awl. Examination of sections, in which the splits retain the stain, reveals that the direction of splitting is determined by the local fibrous structure, being parallel to the fibre bundles and to the osteones. The fibrous structure in the deeper layers of the compacta is ascertained by the same method, after removal of the superficial layers by decalcification. In long bones it is found that the course both of the fibre bundles of the general lamellae and of the osteones is, in general, parallel to the long axis of the bone, but there are deviations from this arrangement. Thus, in rugosities for tendon attachment, the awl makes either round holes or irregular splits, because here the fibres are arranged at greater or lesser angles to the surface, while at joints the holes are circular, because of the vertical arrangement of fibrils in the calcified cartilage. The fibrous architecture revealed by this method in the compacta of flat bones is more complicated, and may be seen by reference to Benninghoff's papers. In general, Benninghoff finds in both long and flat bones that the osteone arrangement of the compact bone is in harmony with, and is a continuation of, the trabeculae of the cancellous bone. This appears particularly well in Benninghoff's diagram (1927) of the radius of the dolphin (fig. 23 B), in which he mapped the osteone arrangement in the compacta (presumably on the face of a section, but this is not clear from the paper) and found a continuation of the structure seen in the spongiosa. He points out how the same trajectorial lines as are "Insubstantiated" in the trabeculae of the

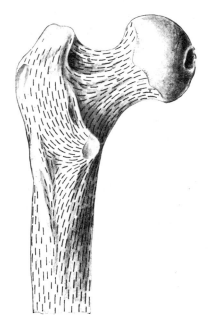

Fig. 26. Split lines in the proximal end of the
human femur. (Henckel, 1931.)

cancellous bone are continued by the osteone arrangement of the compacta and attributes the compact texture of the bone in this region to the condensation of the trajectorial lines into a smaller space, so that they run close together. Henckel (1931) disagrees with this. He points to a number of instances in which the architecture of the compacta, as indicated by its lines of splitting, cannot be regarded as a condensed continuation of the spongiosa architecture. Thus, in the human femur, fig. 26 illustrates his assertion that there is in the proximal end little resemblance between the two structures. Again, he claims that there are cases where homologous bones of different species show similar lines of splitting but dissimilar arrangements of the spongiosa; such are, according to Henckel, the femora and hip bones of the Chimpanzee and Man.

At its lower end the femur enlarges to form the two condyles (fig. 28 A and B). The first figure shows that, as the thickness of the diaphysis increases, its stresses are taken up by parallel, vertical elements of the spongiosa which carry them to the whole of the articular surface. The second figure reveals that these elements are actually perforated plates of bone extending from the trochlear to the condylar surfaces. In spite of the slight outward rotation of the lower leg at the end of extension and the compensating internal rotation during flexion, the knee is in the main a hinge joint, the tibia moving on the femur in a single plane. The joint is under stress whatever the position of the condyles with respect to the tibia.

The pressure is transmitted from femur to tibia through whatever point on the condyles happens to be in contact with the tibia. This point is on the anterior part of the condyles if the knee is straight, but moves posteriorly if the knee is flexed. Thus any pressure trajectory, transmitted through the diaphysial compacta of the femur into the distal spongiosa, may be thought of as swinging through an angle in a plane parallel to that of the movement of the bones (fig. 27 A). In the transmission of the stress, the expanded anterior and posterior aspects of the condyles are more important than their lateral sides, for, in the flexed condition, the femur + tibia represents a bent beam under pressure in a direction tending to increase the bend, and the pressure trajectories are concentrated on the concave, posterior aspect, those of tension on the convex, anterior aspect. The stresses transmitted through the spongiosa may be thought of as a series of lines, indicated in the sectional diagram (fig. 27 B) by the rows of dots (1, 2, 3, etc.). With the joint extended the lines will pass

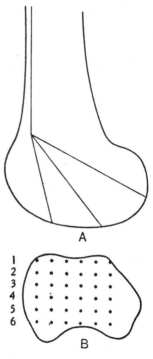

Fig. 27. A and B. Explanation in text.

through the section (assuming them to have been transmitted through the femoral diaphysis on its anterior side) in the position 1, but as the joint is flexed they will come progressively to occupy the positions indicated by the rows 2, 3, etc. The most suitable method of meeting such stresses

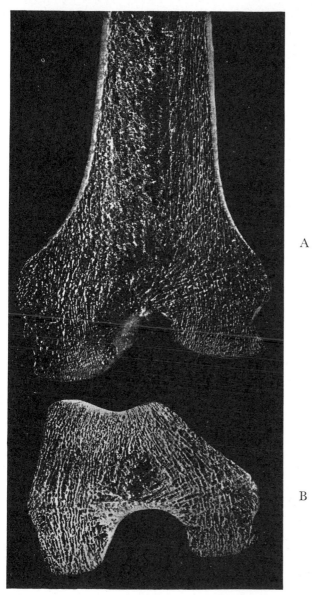

A

B

Fig. 28. A, frontal, and B, transverse sections of the distal
end of the human femur. (Jansen, 1920.)

is to build a series of vertical plates, such as are shown to exist in the condyles by fig. 28 A and B. In view of the fact that the stresses do not actually exist as discrete vertical lines, but would rather impinge on the section in fig. 27 B as a line through each of the series of dots 1, 2, 3, etc., the only more efficient architecture would be solid condyles, but the organism has abandoned solid condyles as being uneconomic and unnecessarily heavy and strong. It should be noted that this discussion requires the plates to run antero-posteriorly, while fig. 28 B shows that they are actually inclined to these planes; this slight obliquity Jansen (1920) attributes to the force exerted by the extensor muscles. An alternative view might perhaps relate it to the small rotation mentioned above.

The structure of the spongiosa at the upper end of the femur is more complicated; its trajectorial "explanation" is usually based on Culmann's comparison with a crane. A diagram of the crane is given in fig. 29 A, a photograph of a frontal section of a femur in fig. 30, and a diagram of the trabecular structure of a femur in fig. 29 B. In the crane, the concave side is under pressure, the convex under tension, and, as in the attached beam (fig. 21) the pressure trajectories start on the tension side but run down the pressure side, while those of tension start on the pressure side and run down the tension side, the two sets of trajectories crossing each other at right angles. Fig. 30, and the diagram fig. 29 B, reveal a close similarity between the crane and the femur, both in general form and in the arrangements of trabeculae in the femur and of the trajectories in the diagram of the crane. The femur may be regarded as a crane to which a projection (the great trochanter) has been attached on the convex side (Dixon, 1910). The trabeculae *aa'*, *bb'* in fig. 29 B carry the pressure stresses, while *cc'* and *dd'* carry tension stresses. This inter-

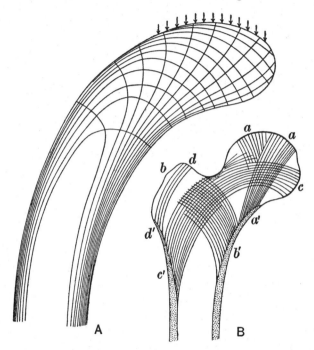

Fig. 29. A. Diagram of trajectories in a crane of the general form of the proximal end of the human femur. (Culmann from Biedermann.) B. Diagram of trabeculae in the proximal end of the human femur. *aa'* and *bb'* = pressure trabeculae. *cc'* and *dd'* = tension trabeculae. (After Meyer).

pretation is supported by the statement that the two sets of trabeculae cross at right angles, but it will be seen that other interpretations have been advanced.

An interesting contrast is drawn by Jansen (1920) be-

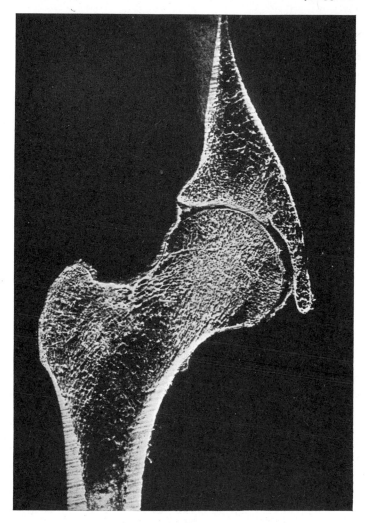

Fig. 30. Frontal section of the proximal end of the human femur, with the acetabulum. (Jansen, 1920.)

tween the structure of the femoral head and of the condyles. While the knee is monaxial, the hip-joint is a ball-
and-socket joint, movements occurring in all possible
planes, and in the femoral head there is no such development of parallel plates as occurs in the condyles.

B. *Criticism of the Trajectorial Theory*

The trajectorial theory has not been without its critics,
and prominent among these is Triepel (1922 a). He cites
twenty objections to the theory; these are of very unequal
weight, and not all will here be mentioned. He points out
that trajectories are lines which are drawn from a number
of more or less arbitrarily selected points a certain distance
apart from each other. If all the possible lines were drawn,
one would arise from every point on the periphery of the
diagram and the latter would be completely blackened;
for obviously the number of possible trajectories is infinite.
Similarly, in a bone, if the trajectories and nothing else
determined spongiosa structure, this would be replaced by
solid, compact bone. Thus something is required to account
for the existence of a lattice structure at all; the trajectories
may influence or determine the architecture of the lattice
when something else has determined that the structure
shall be lattice-like and not solid. Secondly, a difficulty of
fundamental importance arises from the fact that the
engineer, in calculating the distribution of trajectories in
a stressed body, assumes the body to be homogeneous in
structure. But a developing bone is not such a homogeneous
body, for, considered as a whole, it consists of hard bone, developing uncalcified bone (osteoid) and soft tissue (marrow);

further, the hard bone itself is not homogeneous, but has a complicated structure at the histological level, consisting of osteones or Haversian systems, contour lamellae, cemented junctions between lamellar systems, and so forth. Thus in Triepel's opinion one is not justified in assuming that the course of the trajectories in bone will be closely similar to that in an ideal homogeneous body. This is answered by Petersen (1927), who contends that cancellous bone does not usually consist of complete Haversian systems, unbroken contour lamellae, etc., but is a breccia made up of innumerable fragments of these structures. This breccia is the result of the never-ceasing process of resorption and deposition by which the bone grows, and its result is a conglomerate whose fragmentary components are of such a variety of form, size, and orientation, that the statistical result of the chaotic anisotropies is equivalent to homogeneity. Triepel admits that if the trabeculae do in fact correspond with the calculated trajectories, it must be accepted that the trajectories are in them materialised; but he denies that the correspondence does in fact exist. Perhaps the most convincing argument on this matter is on the question of orthogonality. All authors admit that if the trabeculae do not cross each other at right angles they are not trajectorial. Triepel complains (and Jansen, 1920, agrees) that trabeculae frequently do not cross at right angles; and that this is so can be seen by anyone who will examine a number of sections of bones, such as those illustrated in Jansen's book. Thus, in so far as in any bone the elements of the cancellous tissue do not cross one another at right angles, the structure is not trajectorial, and some other principle must be found. Another point concerns the irregular form of the elements; to be trajectorial each trabecula must be a regular curve or a straight line, whereas in reality the trabeculae have not this form, but are often definitely and irregularly, though slightly, bent, and further have many irregular edges with indentations and projections which do not fit a trajectorial scheme. Where two trabeculae cross, the angle between them should be a sharp right angle, whereas actually corners are rounded off, so that a space enclosed between four trabeculae is not rectangular but circular or elliptical. The mid-lines of the trabeculae may indeed cross one another at right angles, and where this is so, the mid-lines may be trajectorial, but some other principle must account for the bony material which rounds off the corners.

In a trajectorial scheme the pressure stresses are crossed by tension stresses, and in the trajectorial interpretation of cancellous tissue it is always assumed that, where an obvious system of pressure trabeculae is crossed by a secondary system, the secondary trabeculae are subjected to tension. Triepel asserts, however, that the secondary trabeculae are not in fact subjected to a simple tension, but to complicated bending moments, because of their attachments to other points in the cancellous tissue and to the compacta. Thus the relations between spongiosa trabeculae and calculated trajectories of stress would be less simple than they appear.

Another point of great interest is Triepel's statement that there exist groups of bones of similar external form but subjected to dissimilar stressing ("Beanspruchung") which nevertheless have a similar internal structure. This of course suggests that the internal structure depends rather on the external form than on the mode of stressing.

The particular group of bones which he cites is the vertebrae, both those of Man from different parts of the vertebral column, and, more remarkably, those of Man and of quadrupeds, which apparently have similar structures in spite of the difference in gait. This would be very convincing were it not for the fact, admitted by Triepel, that the dorsal musculature is similar in all cases, so that the mode of stressing may not be so different after all. It is nevertheless difficult to believe that the mechanical conditions in a human vertebra and that of a quadruped can be at all closely similar.

Triepel concludes that the spongiosa elements are not insubstantiated trajectories, but does not deny that mechanical stresses "stand in causal relation to their development and modification" ("Ausbau" and "Umbau"). He then proceeds to the development of a new principle, which will be described below.

Jansen (1920) and Carey (1929) are in agreement with Triepel in denying the trajectorial nature of bone structure, but differ from him in refusing to admit that tension stresses play any part (save a negative one) in determining its architecture, while holding that pressure is all-important. These authors regard such a bone as the human femur as a pressure-resistant structure whose cancellous architecture is related to the pressures resulting from the weight of the body but especially to the back pressures exerted by the contraction of muscles. Culmann and Meyer interpreted the architecture in terms of resistance to body weight only, not considering the action of muscle at all. But in fact the force exerted by muscle action may greatly exceed the dead weight of the body. Thus Christen (quoted from

Jansen) states that in standing on one foot and on tip-toe, the weight of the body (60 kg.) is transmitted to the ground through the metatarsals, but the quadriceps, in order to balance this and maintain the position, must exert a force of 240 kg., and this is transmitted through the tibia. That muscular force is often greater than body weight is shown by our ability to jump, which would otherwise be impossible, and is impossible in such heavy animals as elephants. In their examination of the frontal section of the femur these authors find the "tension" tracts (fig. 29 B, cc' and dd') of the lateral side of the upper end to be in reality "back pressure vectors" (Carey) of the pull exerted by the glutei and by the short pelvo-femoral muscles which actively press the femoral head into the acetabulum. Triepel is at least in general agreement in considering that the adherents of the trajectorial theory have been led into error by their failure to consider the action of muscles.

Jansen and Carey certainly show that many, if not all, tracts of cancellous bone which have in the past been regarded as tension tracts may be, and probably are, in reality subjected to pressure. In discussing the secondary (transverse) elements of cancellous bone, they are in agreement with Triepel in finding many non-rectangular crossings, and Jansen points out, again with Triepel, that these elements are not necessarily subjected to tension when they cross pressure-bearing systems, because the primary pressure-bearing trabeculae may not be straight, but wavy. In the diagram (fig. 31) it is clear that if the vertical lines are subjected to pressure the cross lines will be under either pressure or tension according to whether they join opposite convex or opposite concave sides of vertical lines. Thus it is not permissible to regard the cross trabeculae as necessarily bearing the opposite kind of stress from that borne by the primaries. In bones whose structure has suffered modification, Jansen finds more evidence against tension as an effective agent in determining bone architecture (see p. 134).

Fig. 31. Explanation in text.

Jansen and Carey are led by their denial of any effective action of tension to a denial of the trajectorial theory. This is justified, for even if the main tracts of cancellous bone be regarded as the embodiments of pressure trajectories, the tension trajectories remain unrepresented and the secondary transverse elements are themselves pressure-resistant structures which take up the transverse pressures caused by the tendency of the primary elements to bend. Cancellous bone on this view is thus not a system of materialised trajectories, but a pressure-resistant lattice, only the primary elements of which in any way represent the trajectories in an imaginary homogeneous body.

Weidenreich (1923, 1924), however, has put forward evidence which makes it very difficult to agree with Jansen and Carey in rejecting tension as a factor affecting the structure of bone. It is found in a number of animals that ossification occurs in tendons, as in certain leg tendons of birds, in the tails of kangaroos, in the dorsal regions of dinosaurs (*Trachodon*, Broili, 1922) and in pterosaurs. The ossified tendons of birds, studied in detail by Weidenreich, show, in section, three zones: an outer zone of unaltered tendon tissue, a middle zone of coarse fibrous (non-lamellar) bone, which Weidenreich regards as a calcification of the tendon tissue without the intervention of specific osteoblasts, and an inner zone of fine fibrous lamellar bone with osteoblasts, Haversian systems, and marrow spaces. The lamellar zone is produced as usual by resorption and replacement of the coarse fibrous bone. In *Trachodon* the ossified tendons, or, in Broili's opinion, more correctly ossified muscles, again have the structure of true lamellar bone, showing typical Haversian systems without general lamellae and with very few interstitial lamellae. In the kangaroo *Macropus* conditions are particularly interesting, for, according to W. Koch (1926–27), the tendons, which are ossified only in a restricted region on the ventral side of the tail, give to that organ the physiological character of a foot. Kangaroos do, in fact, use their tails as third legs for purposes of support though of course not for propulsion, as well as for balance when hopping. The tendons are attached on the ventral side to the 15th–19th tail vertebrae. Ossification does not extend over the whole length of the tendons but is absent over a stretch at both proximal and distal ends, that is at the bone and muscle junctions. Furthermore, the ossification in each tendon is broken at intervals of 4–15 mm., the breaks being scarcely $\frac{1}{3}$ mm. wide; thus the ossified region is not completely rigid but has a very limited mobility. Histologically the bone is of the lamellar type, consisting of Haversian systems, while inner and outer general lamellae are absent and interstitial lamellae rare. The bone is in the form of a tube, the middle region of the tendon not being ossified; the nature of the tissue occupying the cavity is not known, having become unrecognisable in Koch's specimen through maceration. Koch, as stated above, interprets the whole structure of the tail in this region as functionally a foot, for the skin is here also modified, in such a way as to resemble the pads on the feet of many mammals, while the ossified tendons provide a nearly rigid plate corresponding to metatarsals and phalanges. He regards the vertebrae as for this purpose insufficient because of their too great mobility.

Weidenreich regards tendon ossification as developed in direct response to tension, and he cites instances in which abnormally high tensions have been accompanied by ossifications in tendons and other fibrous structures which are normally not ossified. One or two of these are mentioned later (p. 153). The evidence as it stands is in favour of Weidenreich's opinion, but he recognises that the mechanical factor is not the only one operating. This is shown by several facts. At least in birds the ossifications do not appear till fairly late in life, being absent in young birds, in which nevertheless the tension in the tendons must be as great as in the old. In the case of the more atypical tendon ossifications, Weidenreich states that there is a general sexual (male) disposition and a special individual one, for identical conditions may produce the tendon ossifications only in a few individuals. In the special case of the tendons in the kangaroo's tail, it is not impossible that the pressure associated with the use of the tail as a third foot may be the effective agent and not the tension exerted by the muscles. It is, finally, not known whether the ossifications should be regarded as special local differentiations or as extensions of the bone to which the tendon is attached; this point will be mentioned again. It is also not known whether in the normal cases the tendons would ossify if the muscular tension were removed. But in spite of these considerations, the facts make impossible an immediate acceptance of the attempt by Jansen and Carey to exclude tension as a factor in the determination of the architecture of bone.

C. *"Harmonische Einfügung"*

Jansen and Carey accept the proposition that pressure is the sole, or at any rate the main, factor in determining spongiosa architecture, but Triepel (1922 *a*, *b*), who does not deny the efficacy of tension, brings forward a new conception. He regards the architecture as primarily dependent on the form of the element, and as only secondarily modified in accord with the functional demands. The mode of dependency of structure on form is expressed in his term "harmonische Einfügung", an almost untranslatable phrase which indicates that the structure is fitted to, or in harmony with, the whole of which it is a part. Thus an internal structure consisting of concentric shell-like spheres united by radiating rods would be "harmonisch eingefügt" in a spherical body, and Triepel considers that cancellous architecture harmonises with the external form of the bones in this sort of way. The relationship between structure and form is geometric, and, writes Triepel, is sometimes quite obvious to the eye, in other cases a suspicion entertained by the anatomist that it exists requires confirmation by the mathematician, while in still other cases even this help fails and one must fall back (for diagnosis) upon that pleasure to the eye which is afforded by a structure of regularly arranged elements. "Wir sind daher berechtigt, eine Struktur dann als harmonisch eingefügt zu bezeichnen, wenn uns ihr Anblick eine ästhetische Befriedigung gewährt." Unfortunately it seems probable that a structure which was developed wholly in relation to mechanical forces, without any direct dependence upon the external form of the bone, might afford a similar aesthetic satisfaction. Also, the arrangement of trajectories necessarily depends upon the form of the bone, and could therefore hardly produce an architecture obviously out of harmony with that form.

Since structure is to depend on form, it follows that bones of similar form should have similar structures, whether the kind of stressing is the same or not. The vertebrae have been mentioned as, according to Triepel, a group of bones in which this is true. In the long bones of Man, Triepel finds a structure which he reduces to its simplest terms in the manner shown in the diagrammatic longitudinal section (fig. 32). In the section, plates of bone, which are, of course, in reality much perforated, arise obliquely from the compacta on each side and pass in a curve, whose convexity is directed towards the epiphysis, to meet the compacta of the opposite side at right angles, meeting and crossing the corresponding plate of the opposite side in the mid-line of the bone again at right angles. If this structure be thought of in three dimensions, it evidently consists of a series of domes ("Kuppel"), *abc* in the diagram, and of calyces ("Kelche"), *dbe* in the diagram, these having

reflected edges and sitting on the domes at the points made
by their apices. One might think of the system dome +
calyx as formed by the solid revolution of an obliquely
placed, curved line, with its convexity towards the epiphysis, and cutting the axis of the bone at 45 degrees.

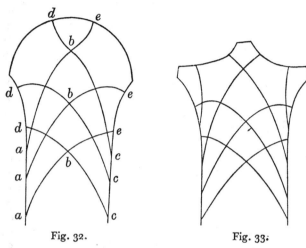

Fig. 32.   Fig. 33.

Fig. 32. To illustrate Triepel's conception of the "harmonisch eingefügt"
structure of the long bones of Man. *abc*, indicate "domes"; *dbe*,
"calyces". (Triepel, 1922 *a*, my lettering.)

Fig. 33. To illustrate Triepel's conception of the "harmonisch eingefügt"
structure of the proximal end of the human tibia. (Triepel, 1922 *a*.)

Successive domes partly enclose one another, and so, of
course, do successive calyces. The walls of the domes arise
more steeply (more nearly tangentially) the nearer the
circle of origin to the end of the bone.

In reality there are many deviations from this scheme.
Thus, some sectors of domes may be more strongly developed than others, and the form of domes clearly changes

in different parts of the bone, being higher (because of the steeper origin of the walls) near the end of the bone. Those arising nearest the end will have no apices, their walls meeting the end wall of the bone, and, especially in bones whose ends are much expanded, become hollow cylinders with parallel or even diverging walls. This is shown in fig. 33, a diagram of the proximal end of the tibia. Other deviations from the simple scheme are produced by other variations of the external form. Thus if the end of the diaphysis is not circular the domes are similarly distorted. Different sectors of dome bases may arise from different levels on the compacta, or at the same levels but at different angles, the consequence being that the apices of the domes lie out of the axis of the bone.

The system of plates which forms the domes and calyces is, of course, completed by other elements, such as horizontal plates, plates in radial arrangement like the leaves of a book (Albert's radiants), and so on.

It is next of interest to see how Triepel interprets the actual structure of a bone which we have already discussed in the light of the mechanical theory. He gives in his book (1922c) detailed descriptions of the spongiosa structure of all or most human bones, studied in sections of a great variety of orientations, whereas Culmann and Meyer and the other trajectorialists have relied mainly upon the structure of the femur, and of this (with the exception of J. C. Koch) almost entirely on the frontal section. Thus Triepel starts with a much fuller knowledge of the material. He regards the proximal end of the femur as containing three systems of domes and calyces which are modified by the form of the bone and the functional demands made upon it. The domes of the first system arise from the compacta at the level of the lower end of the small trochanter and in slightly curved lines like a screw; this spiral structure was also observed by Dixon (1910). The uppermost apices of the domes lie immediately under the compacta on the upper side of the neck near the great trochanter, the calyx walls of the medial side passing into the neck and head, those of the lateral side losing themselves in the great trochanter. In the neck and head lies the second series of domes and calyces; the upper walls of the domes are formed by the medial walls of the upper calyces of the first system, the lower walls by the strong series of elements that arise from the diaphysial compacta on the lower side of the neck and pass to the upper side

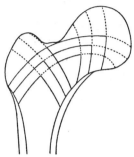

Fig. 34. Illustrating Triepel's conception of the "harmonisch eingefügt" structure of the proximal end of the human femur. "Domes" in continuous lines, "calyces" dotted except when they also form parts of "domes".

of the head, the main series of pressure elements on the trajectorial theory (fig. 29 B, *aa'*). This second series of domes extends into the femoral head, where the last calyces enclose the attachment of the ligamentum teres. I have tried to illustrate this conception in fig. 34. In this frontal section the "pressure lines" of the trajectorial theory are the sectioned lateral calyx and medial dome walls of the first and second systems, while the "tension lines" are the lateral dome and medial calyx walls of the first system, and also the lateral (upper) dome and medial (lower) calyx walls of the second. In addition, there is a third system in the great trochanter and various complications, such as two radiants. Full descriptions are to be found in Triepel's book. To mention another example, a comparison of fig. 23 (facing p. 100) with fig. 32 will show how admirably Triepel's ideal scheme fits the structure of the long bones of the dolphin's fin; that the agreement with the trajectorial theory is equally good illustrates the difficulties of the subject!

Triepel's general position may thus be summarised: fundamentally the internal structure of bone depends upon the external form, to which it is "harmonisch eingefügt". There is for each general form an ideal structure, but this ideal structure suffers modification in accord with changes in the form of the bone and in accord with functional demands. The mechanical conditions are thus not the main factors in determining the architecture, but are secondary factors which modify the ideal structure which, but for them, would be formed wholly in accord with the geometrical relationship.

It is not easy to set a value upon such a conception as this. The idea of "harmonische Einfügung" is attractive, if rather vague, for it is in accord with the general harmony between an organism and its parts; and the notion of existing structure as a compromise between an ideal on the one hand and one which would be truly trajectorial on the other suggests in a general way a line of explanation of the difficulties with which a purely trajectorial theory finds itself confronted. But one consideration weighs very heavily against it. This is derived from the facts of development and has been mentioned in chapter II, pp. 48 and 50; the failure of the geometrically determined ideal structures to appear before the onset of function, the very time when one would expect them to be most obvious, and their actual development only when one would expect the structures to be functionally determined. On the other hand, instances have been mentioned (pp. 31–38) in which structure clearly was determined, not indeed by a mere geometrical relation with the external form, but by the influence of form upon the results of a particular mode of bone development. This is one of at least two ways in which the form of a bone may determine its structure; the other lies in the fact that the course of trajectories through a bone necessarily depends on its shape. The last two points emphasise the need for remembering that form does influence structure, but, owing to the awkward facts from the development of bones, cannot cause us to accept Triepel's interesting conception as it stands.

(2) THE MODIFICATION OF BONE STRUCTURE

That bone structure can undergo very considerable changes following alterations in its mode of stressing is well known, and Wolff (1892), in his classical work *Das Gesetz der Transformation der Knochen*, extended the trajectorial theory to cover these modifications. From a study of a very large number of specimens, he concluded that the structure of the bones, originally the embodiment of the trajectories of force produced by the normal mode of stressing, became, by more or less drastic modification, a similar embodiment of the new trajectorial system.

To this it is objected by Triepel (1904, 1922 *a*) that every such pathological specimen is mechanically of great complexity, and that the distribution of trajectories in it can only be understood after a thorough mechanical analysis. The objection means, in effect, that Wolff's trajectorial interpretations are unacceptable because the modes of stressing of the bones are not sufficiently well known, and because, even if this were known, the courses which the trajectories would follow in homogeneous bodies of the required forms are not known. He points out that only one modified bone has been subjected to a thorough mechanical analysis. This is the case of a knee-joint ankylosis, studied by Roux (1885) in a well-known paper; an examination of this, the classical case and the most thoroughly studied of all bony modifications, will show that Triepel's criticism is not without reason.

The specimen consisted of the knee-joint, ankylosed in a flexion of about 80 degrees; femur and tibia are in bony union in the region of the inner condyle. Except for short lengths above and below the ankylosis, the greater part of both tibia and femur had evidently been sawn off at some earlier date. In addition, the specimen had been partly dissected, the compact bone having been removed on the inner (ankylosed) side, and with it a part of the spongiosa.

We may begin with Roux's assumption that the concave side was the side subjected to pressure by the weight of the body in walking; and the question arises whether in fact the leg was so used. The case history seems to be unknown; but it is at least unlikely that its use in walking (except perhaps by means of some device attached to the knee and continuing the general line of the femur) can have been possible, considering the unchangeable flexion of 80 degrees at the knee-joint. But if the leg was not used in walking, the body weight was not transmitted through it, and the only weight acting on the knee-joint was that of the foot and shank which, of course, would tend to extend the knee, setting up pressure in the convex side, tension in the concave.

The first point of interest in the structure of the specimen is a modification of the compacta (fig. 35): on the anterior

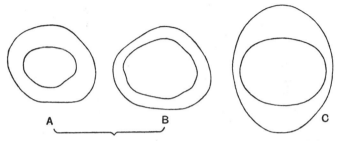

Fig. 35. A and B. Outlines of transverse sections of two normal femora taken 10 cm. above their lower ends. C. Outline of a transverse section through the ankylosis; note the thickening of the compacta on anterior and posterior sides. (Roux, 1885.)

and posterior aspects it is much thickened, while on the sides it is thinner than in the normal bone. Both femur and tibia are modified in this manner.

In his study of the cancellous architecture, Roux distinguished three sets of structures, and obtained a trajectorial diagram of the first structure by an ingenious method. He made a rubber model of the specimen and covered its surface with molten wax or stearic acid; when the model was subjected to bending in the manner in which Roux considered the ankylosis to be stressed, cracks appeared in the wax or stearic acid and from these was inferred the direction of the trajectories set up in the model.

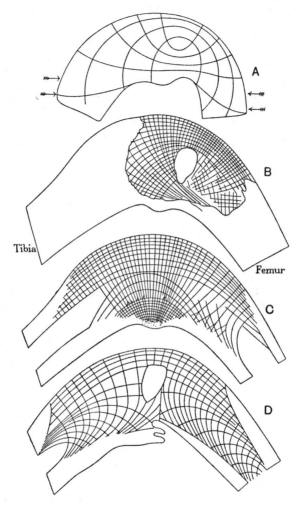

Fig. 36. A. Trajectorial diagram of Roux's model. B. The ankylosis, diagram of first set of structures, revealed by dissection on the medial side. C. The second set of structures, a diagram constructed from several sections. D. The third set of structures, near the outer surface, where union was incomplete. (Roux, 1885.)

The trajectorial map thus obtained is shown in fig. 36 A, and Roux considered that it closely resembled the first set of structures shown in his diagram in fig. 36 B. This first set of structures was made visible on the medial aspect by the removal of the compacta, at least partly before Roux's examination. The second set (fig. 36 C) is seen in sagittal sections, the third (fig. 36 D) in a section parallel to this but situated more towards the outer side, where the ankylosis was incomplete. Roux considers that the first set of structures consists of two sets of trabeculae, pressure- and tension-resistant respectively, each set beginning and ending on one side. Jansen has severely criticised this interpretation, on the ground that such a structure is shown neither by a photograph of the specimen nor by Roux's diagram of the structure (fig. 36 B). In both photograph and drawing the "pressure" trabeculae which start from the anterior aspect of the tibia cross over to the posterior aspect of the femur, but do not return to the anterior side, as they must do if they are to correspond with Roux's trajectorial diagram (fig. 36 A); the same, according to Jansen, applies to the "tension" tracts which he says do not in fact start on the concave side of the tibia, cross over, and return to the same side of the femur. Roux recognises this deficiency in the case of the "pressure" trabeculae, but states that the arc is shown as complete in a drawing made twelve years before by Köster, at a time when a smaller amount of material had been removed. This drawing Roux does not reproduce "weil sie in der speziellen Linienführung eine Anzahl, im Sinne des Zeichners kleiner aber für die mathematische Natur der Verhältnisse schon erheblicher Abweichungen enthält"! Neither does he show a photograph of the preparation, which, however, may be seen in Jansen's book, reproduced from Wolff's (1892). I think, however, that it can be shown that the presence of completed arcs is not so important as might appear, and this for two reasons. Firstly, the model and the trajectorial diagram derived from it represent a short curved structure of the same size as the specimen, and the trajectories drawn in it are without doubt those existing when such an object is subjected to bending in the manner shown by the arrows. But, after all, the specimen was once part of a complete individual, and therefore the two arms of the model should be as long as the whole femur and tibia respectively. There will now be trajectories arising from the two sides at the ends and along the whole lengths of the two arms. Trajectories arising near the remote ends of the femur and tibia will cross the neutral layer near their points of origin, for only so in a structure of this form can they cross it at 45 degrees, then they run along close to the surfaces of the bone and pass round the region of the bend, of course without again crossing the neutral layer. In the region of the bend the trajectories will run more closely together near the surfaces than near the neutral layer, expressing the fact that the intensity of stress increases with the distance from the neutral layer. Thus in the region of the bend the most intense forces are exerted, not across the neutral layer and back again, but straight along the convex and concave sides and near them, without crossing from one side of the structure to the other. Actually, fig. 36 B shows that the main trabeculae of the convex side do in fact behave in this manner, passing straight down the convex side without crossing to the other side. Conditions on the concave side are more complicated, but it must be remembered that this structure lies near the surface on the medial side of the ankylosis, that the architecture can therefore not be expected to show the simplicity of a sagittal section, and, finally, but perhaps most important, that the specimen had a double curvature. It was concave not only posteriorly but to a less extent medially. There were therefore two sets of bending stresses and the minor set would be expected to modify the consequences of the major set especially on the medial side. The second set of structures (fig. 36 C), that in the sagittal plane, is interpreted by Roux in a manner which, save in being more detailed, corresponds well with these considerations; there are two main masses of trabeculae, on the anterior and posterior aspects respectively, crossed at right angles by trabeculae running from the anterior side to the posterior. The latter trabeculae probably resist the pressure

Fig. 37.

which tends to bring anterior and posterior sides together when any force is exerted tending to bend the ankylosis, just as opposite sides of a piece of straw come together when the straw is bent (fig. 37). According to Jansen, such trabeculae, running straight across from one side to the other, are always present in ankyloses fixed in any position other than extension. It seems not impossible that the trabeculae shown in fig. 36 B, passing from the anterior to the posterior sides, may be of the same nature, i.e. not arc trabeculae as Roux supposed, but struts keeping the anterior and posterior sides apart.

The third type of structure exists on the lateral side,

where the ankylosis is not complete, the bones being united only towards the posterior side. Roux regards the architecture shown in fig. 36 D as a resultant of two factors: the pressure between the anterior and posterior sides (which was regarded as producing the antero-posterior trabeculae in the second set of structures) and that between the united surfaces of the two bones, while the trabeculae running along the convex side, in the region of union, of course correspond with those occupying similar positions in the two other types of structures. It may, however, be noted, as an alternative, that the structure shown is roughly what one would expect at the ends of two long bones each separately subjected to bending, i.e. at the region of the ankylosis if the two arms were not, or only very incompletely, united, and each subjected to bending forces about the middle of its length.

This ankylosis is the modified bone which has been most thoroughly examined, and the discussion will have shown that our understanding of even this case is far from complete. Yet certain conclusions emerge: the original form and structure have suffered almost complete destruction and been replaced by new, and, while Jansen's criticism of the details of Roux's interpretation seem to be justified, nevertheless, if greater significance is attached to the simpler and better known arrangement of bony systems in the sagittal section, and if it be remembered that the specimen was, after all, once part of a whole man, the structure appears suited to a single bone subjected to bending. This, indeed, Jansen does not deny; he is merely concerned to eliminate tension stresses as a factor in the production of the architecture. But this and many other

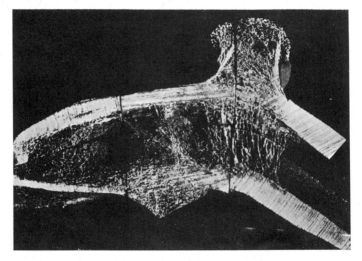

Fig. 38. Knee-joint ankylosis. (Jansen, 1920.)

ankyloses show strong bony tracts on both convex and concave sides. It is usually considered that both sides cannot be subjected to pressure; to this Jansen replies that either side may be under pressure at different times, and for this he has four reasons: (*a*) pressure can be exerted through the convex side of the ankylosis by the M. rectus femoris, an extensor muscle which, being poly-articular, is maintained in a "relatively good condition" while other muscles suffer disuse-atrophy; (*b*) if the angle of flexure is small, standing with the trunk bent forward will send pressure stresses through the front part of the knee-joint, because the line of gravity then passes in front of the transverse axis of the joint; (*c*) the weight of the foot, if or when it is clear of the ground, tends to extend the joint and so exerts pressure on its anterior side; (*d*) sudden removal of tension in the front part of the joint (as in walking) "may be followed by vibratory movements during which the tension stresses in the front part of the knee alternate periodically with pressure stresses".

Jansen figures a knee-joint ankylosis which is especially interesting because of the drastic nature of the structural changes which it has undergone. His photograph is reproduced in fig. 38. The most striking features are the great thickness of the compacta on both convex and concave sides in the region of the ankylosis, the part which normally is very thin (compare figs. 24, p. 102, and 28 A, facing p. 106), the disappearance of any trace of demarcation between femur and tibia, and the very complete reorganisation of the spongiosa. The chief character of its internal architecture is the strong system of struts running from the anterior to the posterior side. Similar changes have occurred in the hip-joint ankylosis shown in fig. 39, which should be compared with fig. 30, facing p. 108. The compacta in the region of the neck of the head of the femur has been greatly thickened, and the tract of cancellous bone which, in a normal femur, runs from the great trochanter along the upper side of the neck (fig. 29 B, p. 108, part of cc') has disappeared, or, expressed otherwise, has been transformed into compact bone, while that part of it which normally lies in the head itself has disappeared. The tract which normally extends from the upper side of the head to the lower side of the neck (fig. 29 B, aa'), the main pressure trabeculae of the trajectorial theory, is in part more or less unchanged save that it extends right across the pelvis, and is in part converted into or replaced by the thickened compact bone of the neck. The tract in the ankylosed specimen running from the base of the great trochanter across to the opposite side of the bone is apparently the tract which normally occupies roughly the same position, strengthened to prevent the collapse of the tube in this position (fig. 37).

A modification of a different type is seen in fig. 40, a badly set fracture. The marrow cavity is spanned by a trabecular system, in correlation with the displacement of the distal part of the bone, necessitating transmission of the stress, arriving at the region of the fracture along one side of the bone, through both sides. Jansen comments upon the formation of cancellous bone where there is enlargement of the transverse section which transmits the stress, as shown here and in the ends of normal long bones.

The architecture of the compacta has been little studied in modified bones, but it is clear that considerable changes occur. Not only may compact bone be converted into

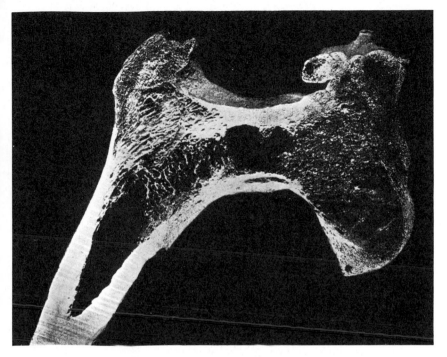

Fig. 39. Hip-joint ankylosis. (Jansen, 1920.)

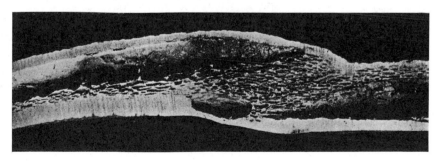

Fig. 40. A fracture. (Jansen, 1920.)

cancellous and *vice versa*, as in several of the cases already described, but considerable alterations probably occur in the architecture of bone which maintains its compact nature. Benninghoff (1927) briefly describes an abnormally curved rachitic femur and Dowgjallo (1932) has studied the structure of the mandible which has undergone the changes associated with the altered mode of functioning after the loss of the teeth. Using Benninghoff's method, he finds alterations in the arrangement of the lines of splitting. These are of such a kind that, for each change of external form, there is a corresponding architectural alteration, the result being always a coincidence between the arrangement of the osteones and the direction of the force acting. Benninghoff's criticism (1934), however, shows that Dowgjallo's description of the normal jaw cannot be unreservedly accepted. He considers that some of the changes which Dowgjallo regards as functionally induced consequences of the changed mode of functioning following the loss of the teeth may be within the range of normal variation of jaws fully equipped with teeth. He agrees with Dowgjallo, however, that the loss of the teeth is associated with atrophy of all osteone tracts especially related to them, but further work will be required to decide whether Dowgjallo is right when he regards the changes as a general reconstruction involving the development of new tracts as well as the loss of the old, or merely an atrophy of parts of the structures which have become useless.

The arguments advanced against the trajectorial theory of modified bone structures need not be discussed again, for they are in the main identical with those put forward in connection with the normal structures. Thus Jansen, in a study of a number of bones whose structures have been in various ways modified, finds that increased pressure always leads to strengthening of the bony structure while increased tension leads to atrophy. In *coxa vara*, a condition in which the head of the femur is depressed so that its angle with the shaft is decreased, the pressure stresses on the concave side are increased and the bone is there thickened, while on the convex side, where the tension is increased, it suffers atrophy. In scoliotic vertebrae he finds it a constant fact that the same occurs, the side of increased pressure being thickened, that of increased tension thinned. He shows similar conditions in a number of other bones whose stressing and structure has been altered. Finally, Jansen concludes with the statement that "The form of the bone being given, the bone elements place or displace themselves in the direction of functional pressure" is the allowable remnant of Wolff's Law.

The question of the efficacy of tension in determining the architecture of bone must remain open. The majority of authors see no distinction in this respect between tension and pressure, but the evidence presented by Jansen and Carey cannot be disregarded. At any rate there is no doubt that the adult, fully formed bone, subjected to a changed pattern of mechanical stressing, can meet the new demands made upon it with a changed structure, and that the new morphological pattern so produced is, in respect of the new mode of stressing, more efficient than the old.

It is well to end this chapter with a brief summary of the position here taken towards the trajectorial theory. It has become very clear that the structure of the spongiosa, the form and thickness of the compacta, and probably the arrangement of its osteones, are in a general way mechanically suited to resist the stresses to which they are subjected. But according to the trajectorial theory the trabeculae of the spongiosa must be regarded as accurate materialisations of the lines of principal force in a homogeneous body of the same form as the bone and stressed in the same way, the compacta must be regarded as formed at least partly by the condensation of the materialised trajectories where they run close together, and the osteones must, according to Benninghoff, be trajectorial materialisations in the same sense as the trabeculae of the spongiosa. Against this extreme view a number of considerations have appeared, such as the absence of orthogonality, irregular forms and wavy course of cancellous trabeculae, the alleged absence of tension trajectories, and the alleged failure of osteone arrangement to harmonise with cancellous architecture. Accordingly, the attitude here taken is that in the determination of bony structure a very large part is played by mechanical forces, that bony structures are therefore in a general way mechanically adaptive, but that the trajectorial theory in its rigid form demands a degree of perfection in this adaptation which does not exist in fact, because the mechanical stresses and strains of function are not the only factors which influence the development of bony architectures.

CHAPTER V

THE MECHANICAL STRUCTURE OF
CARTILAGE

The recognition of a mechanical structure in bone led naturally to the enquiry whether the other principal tissue of the skeleton, cartilage, could similarly be said to have a structure mechanically fitted to resist the stresses to which it is subjected.

The mechanical properties of cartilage are very different from those of bone; of the two, bone is by far the more resistant to deforming forces and offers nearly equally strong opposition to both pressure and tension. Cartilage, on the other hand, is in an absolute sense less resistant than bone, but is more elastic, and is, at least when considered apart from its fibrous sheath, much better able to withstand pressure than tension. Rauber (figures from Kopsch, 1922) found that pieces of cartilage were able to resist tension amounting only to 0·17 kg. per mm.^{2} but pressure up to 1·57 kg. per mm.^{2} According to Benninghoff (1925 b), Triepel (1902) obtained similar results. These figures tell us something of the mechanical properties of fragments of cartilage but nothing of cartilage as a whole, and, as will shortly appear, cartilage differs from bone, not only in the respects mentioned, but also in being a more fully integrated tissue in the sense that the "Festigkeit" of cartilage depends more than that of bone on the entirety of the whole cartilaginous organ. Whereas the mechanical properties of bone are not dependent on those of the periosteum, to separate cartilage from its perichondrium is to disrupt a mechanical unity, for cartilage owes its powers of tensile resistance largely to the fibres of the perichondrium and, probably in a lesser degree, to the fibrils embedded in the matrix. The reasons for the low figures found for tensile resistance by Rauber are without doubt that the perichondrium was not included in the fragments or else was not complete, and that the direction of tension was not known to be that of the collagen fibrils in the matrix. Benninghoff (1925 b) found that in the sub-perichondral zone of tracheal cartilage the resistance to tension acting in the direction of the fibrils was 1·22 kg. per mm.^{2}, seven times the figure found by Rauber, and this without including the perichondrium.

It is interesting that the elasticities of different cartilages are not the same, even between serially homologous cartilages in the same animal. Thus in young rabbits and dogs the first rib cartilage cannot be bent as much as the seventh, although the two are of about the same thickness. The respiratory movements of the seventh rib are greater than those of the first, suggesting that the mechanical properties of the tissue have been developmentally adapted to the forces acting (Benninghoff, 1925 b).

The apparently simple structure of hyaline cartilage, seeming to consist of cells and homogeneous ground substance, reveals itself under critical investigation as very much more complicated than this. The components (figs. 42, 44, facing p. 142) are ground substance, collagen fibrils, and cells, and the ground substance again has an elaborate structure. Each cell is invested by a thin layer of matrix, the capsule, and the capsule is again enclosed in a thicker zone of matrix, typically of more or less spherical form, the whole structure of cell, capsule and wider zone of matrix being known as a *chondrone* or *chondrin globe* or "*territory*". Frequently more than one cell is enclosed in a single chondrone. Between the chondrones, whose material appears to be laminated concentrically around the cell, lies the interterritorial substance filling up all the spaces between the chondrones. The architecture is further complicated by the arrangement of the cells, and therefore of the chondrones, in definite patterns, usually columns, and by the existence, between these columns, of trabeculae of inter-territorial substance which are thicker and more continuous than the shorter trabeculae which cross the columns, separating adjacent chondrones. The collagen fibrillae which run in both chondrones and inter-territorial substance, differ from those of ordinary connective tissue in not being grouped in large bundles, but run independently of one another. They form a complicated structure of interlacing fibrils, but in this it can be detected that the fibrils show certain predominant orientations. We are thus dealing not with a structureless tangle but with an orderly and very complicated architecture. The direction of the fibrils has been worked out partly by ordinary histological means but also by other methods. One of these is the use of polarised light. Another method, more useful for studying the arrangement of fibrils at the surface of a cartilage, is that already mentioned and used by Benninghoff (1925*a*) for the study of the architecture of compact bone, and consists in driving an awl into the cartilage and seeing in which direction the matrix splits.

This fundamental structure of cartilage, in so far as it consists of chondrones and inter-territorial substance, is well enough understood; it is somewhat otherwise with regard to the relation of the included collagen fibrils to the other structural components. It is evident that we cannot hope fully to understand the relation between cartilage structure and function until the fibrillar component has been more adequately mapped. The account which follows is that given by Benninghoff (1925 b), who is at pains to point out that it is diagrammatic and, at least to some extent, provisional. It has, however, the advantage of being in accord with the available facts, of providing a synthetic description applicable to cartilages in general (with the probable exception of those of embryos, in which the fibrils are perhaps absent) and of suggesting a rational connection between structure and mechanical stresses. Benninghoff envisages the fibrils as running concentrically around single cells, around chondrones containing several cells, and around groups of chondrones. The idea is best expressed by the reproduction of his diagram (fig. 41, p. 140). The fibres of the perichondrium form the outermost of many series of concentric systems, in which each cartilage cell, each chondrone, and each group of associated chondrones (like a cartilage column) are all enclosed in investing systems of fibrils.

Up to a point there is no doubt about the part which these various components play in giving the cartilage as a whole its mechanical properties. Fibrils are certainly tension-resistant, and not pressure-resistant, structures; thus tensile stresses must be taken up by the perichondrium and by the fibrils embedded in the inter-territorial substance and chondrones. The non-fibrillar component, as is shown by Rauber's figures, is much less tension than pressure resistant, and therefore resistance to pressure must be performed by the matrix and the cells. Now a chondrone may be compared to a fluid-filled bladder with thick and elastic walls. The fluid content (the cell) prevents decrease in its volume under pressure, but it can be deformed, and,

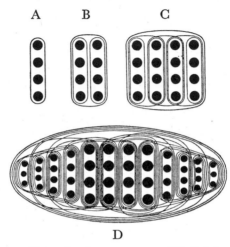

Fig. 41. Diagrams illustrating the arrangement of fibrils in hyaline cartilage. A. A single chondrone. B. Two chondrones. C. A group of several chondrones. D. An entire cartilage. Each black circle indicates a single capsule with its own fibrillar sheath. In D the outer encircling lines represent the perichondrium. (Benninghoff, 1925 b.)

since matrix has little power of tension resistance, would easily be burst, were it not for the concentric sheaths of fibrils embedded in the matrix. Similarly, a group of chondrones, held together by inter-territorial matrix, would very readily be sheared apart if this were not prevented by the systems of tension-resistant fibrils. Considering an entire cartilaginous element, the same principle obviously applies, the perichondrium being the external tension-resistant component which ensures the cohesion of the whole.

Cartilage thus differs from bone in the more obvious histological separation of structures performing different mechanical functions. In bone the pressure-resistant component is the deposit of calcium salts, the tension-resistant the bone fibrils, while the cells play no directly mechanical role. In cartilage the functions of pressure resistance, provision of resiliency, and tension resistance are represented by three different structural bases. The cells, forming the incompressible fluid contents of the chondrones, and to a lesser extent the matrix, provide resistance to pressures, while the compressible but elastic matrix, with its embedded fibrils, ensures that the structure shall have the resiliency which is probably so important in buffering mechanical shocks, while the fibrils themselves, as well as resisting primary tensile stresses especially in the perichondrium, ensure the maintenance of the whole structure by taking up shearing stresses which would otherwise destroy the cohesion of the structure.

These structural-mechanical relationships are well illustrated by an experiment performed by Benninghoff, 1924 (fig. 43, facing p. 142). If the cartilage in a rabbit's pinna is subjected to moderate pressure by bending, the convex side, in which of course the tension is increased, shows a thickening of the perichondrium. On the concave side of the bend, where the pressure is increased, there may be, at the point of maximal pressure, dissolution of cartilage and its replacement by connective tissue, but where the compression is not so great new cartilage is formed from the perichondrium and the character of the old cartilage changes. It loses most of its elastic fibres, and the matrix walls of the chondrones become thicker at the expense of the space occupied by the cell within. The new cartilage forms an inseparable unit with the old and has the same character, with few elastic fibres and many small thick-walled chondrones. Cartilage having this structure is obviously more resistant to the deforming effect of pressure than the unmodified cartilage, which, with its large thin-walled chondrones and rich supply of elastic fibres, remains unchanged on the tension side. Thus there is on each side an increase in those elements adapted to resisting the kind of stress which is locally predominant and, at least on the pressure side, a loss of those elements (the elastic fibres) which are not required. The theory advanced by Benninghoff to account for these changes may be read in the original paper.

As an example of the architecture in a normal cartilaginous organ we may take the tracheal cartilage of cattle (fig. 44). This has the form of an open ring, subjected to bending in the sense of opening and closing the ring under the action of plain muscle. The cartilage, as described by Benninghoff (1925 b), has the following structure: The perichondrium is very thick on the convex side, very thin on the concave side; on the convex side it is firmly attached to the underlying cartilage and cannot be torn off, while on the concave side it can be torn away and only a few fibres are broken in the process. Lying beneath the perichondrium on the convex side is a transitional zone composed of fibres with hyalinisation gradually increasing with depth; many of the fibres can be seen to bend downwards

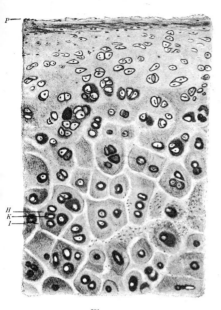

Fig. 42.

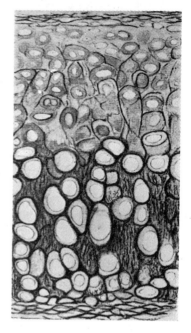

Fig. 43.

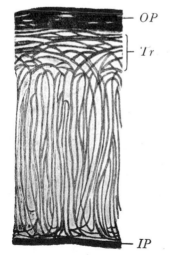

Fig. 44.

Fig. 42. Section of tracheal cartilage of Man. (After Schaffer.) *P* = perichondrium; *K* = capsule; *H* = chondrone; *I* = interterritorial substance.

Fig. 43. Illustrating Benninghoff's experiment on the elastic cartilage of the rabbit's pinna, concave side above. Explanation in text. (Benninghoff, 1924.)

Fig. 44. Diagram of the arrangement of fibrils in the tracheal cartilage of cattle. *OP* = outer perichondrium, *Tr.* = transition zone. *IP* = inner perichondrium. (Benninghoff, 1925 *b*.)

into the true cartilage substance below. Like the peri-chondrium, this transitional zone is much thicker on the convex than on the concave side. Underlying it is the true cartilage. This consists of more or less isodiametric chondrones arranged in columns whose axes run roughly along the radii of the ring and which are separated by trabeculae of inter-territorial substance. The fibres of the perichondrium run for the most part horizontally around the cartilage but many of them dip down into the transition zone. In the latter they run in part horizontally round the cartilage, but large numbers can be seen to bend down into the cartilage substance below, forming loops or arcades. In the cartilage itself they run mainly in the inter-territorial trabeculae between the columns, and are anchored in the transition zone, joining the horizontally running fibres of that zone or forming loops which turn back into the cartilage.

The greater development of the perichondrium and transition zone on the convex side is evidently in accord with the fact that this side is that subjected predominantly to tension, while the displacement of the cartilage substance towards the concave side similarly accords with the pressure to which that side is subjected.

But here a difficulty arises. In a bent beam of homogeneous composition the neutral layer is equidistant between the two surfaces. In an artificial beam made up of two components, one pressure resisting, and the other tension resisting, and both of equal efficiency for their respective tasks, the tension-resisting substance would, of course, be concentrated on the convex side, the pressure-resisting on the concave side, and each would occupy exactly half the thickness of the beam, the neutral layer separating the two components and occupying the same position as in a homogeneous beam. Now, in the cartilage, this position, equidistant between the two surfaces, lies in the zone occupied by the pressure-resisting cartilage substance, the junction between mainly pressure-resistant and mainly tension-resistant zones being in, or just beneath, the transition zone on the convex side, by no means midway between the two surfaces. But Benninghoff remarks, probably correctly, that the amount of deformation which the chondrones can suffer under pressure is greater than the amount by which the fibrils of the perichondrium can extend under tension: in other words, the two systems are not of equal efficiency. Evidently, then, the lower efficiency of one substance must be compensated by its presence in greater quantity, as though in an imaginary beam the neutral layer had been displaced towards the tension side and the enlarged pressure side constructed out of a less efficient material. This would be possible because the trajectories on the pressure side would no longer run so close together, i.e. the intensity of stressing would be decreased.

Benninghoff recognises that pressure trajectories are not represented in this structure. The columns of chondrones and the bars of inter-territorial substance between them have indeed a trabecular arrangement, but the directions of the trabeculae are along the radii of the tracheal ring, not at all in the lines of principal pressure, for the greatest intensity of pressure must be transmitted near and nearly parallel to the surface on the concave side. He does, however, consider that the fibrillar architecture is trajectorial, and, since the greatest intensity of tension must be transmitted near and nearly parallel to the surface on the convex side, this is true of the fibrils in the perichondrium. But in the hyaline cartilage itself, the fibrils are oriented between and around the columns of chondrones, and so, like them, along the radii of the ring. Now, around whatever plane the cartilage be supposed to bend, the tension trajectories must cross the neutral layer at 45 degrees, but, since the neutral layer is parallel to the two surfaces, the fibrils cross it rather at 90 degrees than at 45 degrees. Thus it seems impossible to regard the fibrillar architecture as truly trajectorial, at least in the sense in which the word is always used in the study of functional structures, as embodying directions of principal stress. It is possible that the fibrillar structure can be interpreted better in terms of the fundamental structure of cartilage and of the disposition of its chondrones than in terms of trajectorial diagrams. If the columnar disposition of the chondrones be taken as given, and if Benninghoff's scheme of multiple concentric fibrillar sheaths be remembered, it becomes clear that the fibrillar architecture of the tracheal cartilage must be what in fact it is, and the factors ultimately responsible for it are those which more immediately cause the columnar arrangement of the chondrones. Now, whatever the plane of bending in the tracheal ring as a whole, the pressure stresses run for the most part parallel to the surfaces. Benninghoff has pointed out that the cartilage is not a rigid material like bone and that a cartilage column cannot resist pressure along its long axis (unless it has extraneous support); it is therefore to be expected that the columns will lie at right angles to the general line of the pressure, along the radii, and not parallel to the pressure, as some authors have thought. Thus the fibrillar architecture, whose most general characteristic is inherent in what might be called the "concentric habit", displays the particular features of tracheal cartilage in dependency on the passive orientation of the cartilage columns. The functional significance of its loops is to prevent the matrix being squeezed out on the concave side whenever the cartilage is bent, while that of the perichondrium is to resist the principal tensile stresses, which, of course, run superficially, remote from the neutral layer. Nor need the "concentric habit" be regarded as a mystery. If any more or less spherical

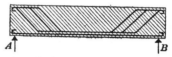

Fig. 45. Diagram of a beam of reinforced concrete.
(Benninghoff, 1925 b.)

bladder-like body, such as a chondrone, be compressed, tangential tension stresses are set up in its walls, and it is well known that fibrils tend to develop along lines of tension. Thus if fibrils are to develop at all in the wall of such a structure they will necessarily be arranged concentrically round it.

Benninghoff has an interesting comparison between the structure of cartilage and that of reinforced concrete (fig. 45). The latter, like cartilage, is composed of two substances, of which the concrete has considerable compressive but little tensile strength, and the steel very great tensile strength. In the diagram of the piece of reinforced concrete it will be seen that the iron rods are placed along the tension side (here the lower), offshoots running up into the pressure side near the ends, so following approximately the lines of the tension trajectories. In comparison with the tracheal cartilage, the iron evidently represents the perichondrium and transition zones, the concrete the cartilage substance. In both structures the appropriate materials are concentrated at the appropriate places. In cartilage the tension-resisting elements are many, in the artificial structures few, and in cartilage they are more intimately attached to the pressure-resisting substance than is the iron to the concrete. The latter is anchored by hooks, the fibrils of cartilage by their innumerable loops and arms in the transition zone and perichondrium.

Heidsieck (1934) has suggested, as an amendment, that a better comparison is with cord motor tyres, in which layers of cord are built in between layers of rubber. The cord is, of course, the tensile component, the rubber the pressure-resistant element, and the comparison is an improvement on that with concrete in that the elastic properties of rubber and cartilage are more alike than those of concrete and cartilage.

Little has so far been said of the factors responsible for the development of cartilage or of the architecture displayed by its components. The first question arising is that of the origin of cartilage itself, and here a negative conclusion at least can be drawn. The development of cartilage in fragments of embryonic limb-bud mesenchyme explanted *in vitro* prior to the onset of chondrification, and some of the cases of ectopic chondrification (e.g. in the kidney, Natanson, 1903), are sufficient to exclude mechanical factors as essential agents in its differentiation. On the other hand, there is no doubt that it can be created anew in adult life, as is shown by such facts as the development of cartilage-covered articular surfaces in pseudarthroses, where the efficient stimulus seems to be the rubbing of two bony surfaces against one another. Another problem is presented by the persistence of parts of cartilaginous organs the great bulk of which is replaced by bone. The replacement by the epiphysial ossification in mammals of the whole of the cartilage except the narrow layer covering the articular surface is presumably related to the absence of calcification except in the basal layer of the future articular cartilage, but that this cannot be the whole story is indicated by the fact that even this basal, calcified region is not replaced by bone. It is at any rate improbable that mechanical factors will be found to supply the answer.

The differentiation between hyaline and the two kinds of fibro-cartilage is the next problem. It is well known that white fibrous cartilage is found where strength is required without the rigidity of bone and elastic cartilage where the tissue must be able to undergo changes of form without rupture or permanent deformation. Here there are two questions which must be indicated, although their discussion will not take us very far: that of the factors evoking differentiation of a particular kind of fibre and that of the mechanism producing arrangement of the fibres. The coincidence between the kinds of fibrous impregnation and the functional needs of the part of course suggest functional factors as the differentiating agency, but a serious difficulty in the way of accepting the proposition lies in the elastic cartilage of the pinna. A pinna supported by white fibrous or hyaline cartilage would doubtless soon be deformed, broken, or torn away, and so it is easily seen why, in a teleological sense, it is well that it should be supported by elastic cartilage. But it is hard to imagine what kind of mechanical stressing could have influenced its differentiation; even the most ferocious of puppies does not chew his brothers' ears *in utero*. Be this as it may, there can be little doubt that the fibrillar architecture of the two varieties of fibrous cartilage is such as to warrant the use of the term "functional structure". A good example is provided by the inter-vertebral fibro-cartilages of Man. These form a fibrous ring around the Nucleus pulposus, said to be a relic of the notochord, which acts as a water cushion between the vertebral bodies. The fibres of this cartilaginous ring are arranged in concentric layers and are in each layer spirally directed, the directions of the winding changing in alternate layers like the fibrils of an osteone. When the vertebral column is bent a complicated series of stresses is of course set up in the inter-vertebral cartilage; pressure is taken up by the Nucleus pulposus and by the matrix and cells of the cartilage, tension of course by the fibrils, and owing to the alternating spirals there is always a fibrillar system suitably oriented. But, just as the mechanism producing white or elastic fibres in cartilage is unknown, so is that which arranges the fibres, and therefore, until some investigator throws light on the problem, further discussion is useless. It is, of course, to be expected that whatever factors influence the orientation of fibres in non-cartilaginous fibrous tissues will be effective in cartilage also, and there can be little doubt that the same mechanism will be found to be at work on the masked fibrils of hyaline cartilage as on the coarser fibrous systems of white and elastic fibro-cartilage.

CHAPTER VI

THE MECHANISM OF BONY ADAPTATION

SECTION A

In much of the literature on bony architectures one finds that no distinction is drawn between the production of bone and its arrangement in certain patterns. It is assumed that, if a bony structure has an architecture which is brought into existence in response to mechanical conditions, the bone must have been produced by the same factors, and therefore that the osteoblast, if it is recognised as a differentiated cell at all, is itself produced in response to the same conditions. But here there is a confusion of thought, for it is surely obvious that osteoblasts may well be created by factors quite other than the mechanical, but the pattern of the structure they produce be determined by the static and dynamic conditions in the region which it occupies.

It is not an object of the present work to discuss the general question of the osteoblast. In the classical theory this cell is regarded as standing in the same relation to bone as the gland cell to its secretion, but some authors consider it to be merely a reaction form of the ordinary connective-tissue fibroblast with no special function in the formation of bone, and others, while agreeing that it is indeed an ordinary fibroblast and not specially differentiated for playing a part in osteogenesis, believe it to be a degenerate form which, in dying, in some manner contributes to the process of calcification. The classical theory is adopted in what follows; but even if it be rejected by the reader the main conclusions can still be accepted, because no special theory of the nature or even the existence of the osteoblast is essential to them.

A large body of facts, some of which will be mentioned, shows that special mechanical conditions are essential neither for the differentiation of osteoblasts nor for the production of bone by osteoblasts that are already differentiated.[1]

H. B. Fell (1932) found that bone was developed *in vitro* by fragments of the periosteum of the long bones of six-day chick embryos. Since other tissues, not normally destined to produce bone, do not ossify under the same conditions, ossification in the periosteal explants must have been due to an intrinsic quality of the explanted periosteum and not to special mechanical conditions acting upon an indifferent tissue. In this experiment osteoblasts were already present in the periosteum when it was explanted, and it therefore tells us nothing of their first differentiation. When earlier embryos were used, and embryonic mesenchyme from the limb-buds of three-day embryos explanted, cartilage was produced but also bone (1928).

In experiments of fundamental importance Huggins (1931, 1933) found that if fragments of the epithelium of the urinary bladder of dogs were grafted into the parietal fascia, rectus abdominus muscle, rectus sheath, subcutaneous fat, fascia lata, or fascia over the sacro-spinalis muscle, the connective tissue around the graft produced

[1] The opponent of the classical theory of the osteoblast may read: "essential neither for the initiation nor for the continuation of osteogenesis."

bone, but when the grafts were made into the parenchyma of the kidney, into the liver, spleen or lung, no bone was formed. Similar epithelial grafts derived from the adrenals, gall-bladder, stomach, jejunum, or colon, the dura mater, and thin shavings of bone, synovia of knee-joint, placed in the rectus sheath, did not induce bone formation. Thus, certain kinds of proliferating epithelium (bladder, ureter, pelvis of kidney) excite bone formation in the connective tissue of certain regions but not in other parts, and Huggins infers the existence of two kinds of "fibrocytes", of which one class ossifies in response to the presence of proliferating bladder tissue, etc. while the other does not. Evidently this is in close agreement with Fell's work described in the last paragraph in excluding special mechanical factors as necessary conditions of ossification.

Many of the other numerous cases of ectopic bone formation may likewise be adduced in the sense of the last conclusion. Thus, the ossification which occurs in the kidney after ligature of its vessels, and which has been described by Sacerdotti and Frattin (1902), by Poscharissky (1905) and by Maximow (1907), that in the connective tissue associated with transplanted guinea-pig ovaries (Voss, 1926), and the ossifications which develop in wound scars in various situations, can hardly be caused by any special local condition of mechanical stress, but must rather be connected with the coincidence of special biochemical conditions and a supply of young, probably proliferating, connective tissue, some of whose cells (on the classical theory) differentiate into osteoblasts.

There are, however, cases in which a mechanical cause of ectopic ossification can be advanced with greater probability. Such are "exercise" and "riding" bones, which develop in particular muscles in association with definite kinds of physical activity, and also ossifications in the walls of arteries, since these are subjected to the intermittent pressure which, as will be shown, is probably the form of mechanical stressing which is most effective in advancing the deposition of bone. In the case of exercise bones, however, it is probable that the responsible cause is not the mechanical stressing itself but the muscle trauma which results from it, and Busse and Blecher (1904) cite three cases of myositis ossificans following a single severe trauma in which the limb was immobilised after the injury. In the case of arterial ossifications, since ossifications can arise in conditions in which no special mechanical factors seem to be operative, it is gratuitous to assume the essential efficacy of such factors when bone happens to appear in a situation in which they are known to exist.

Weidenreich (1923, 1924) is of the opinion that ossification of tendons is a direct response to the tension exerted through them, and it appears to be true that such ossifications tend to appear in tendons and similar structures subjected to heightened mechanical stressing. Thus Weidenreich (1924) mentions the case of a woman of forty-six years who had a congenitally split pelvis; there was no pelvic symphysis, but in spite of this the woman was able to work in the fields, and certain ligaments, which because of the skeletal abnormality were subjected to supernormal tensions, were ossified at their attachments for considerable distances. In another case, demonstrated to Weidenreich by Loeschke, the ligamentum longitudinale anterius was ossified through nearly its whole length and was subjected to an abnormal degree of tension by a lordotic[1] stretching of the anterior vertebral region. The normal ossified tendons of birds, kangaroos, and dinosaurs were discussed in an earlier section. It is impossible to deny that these ossifications are related to the special conditions of stressing to which the parts are subjected. Now, Weidenreich (1923) finds, in the ossified tendons of birds, that the development of lamellar bone is preceded by coarse-fibrous bone which he regards as no more than a calcification of the tendon. Thus, the mechanical stressing may be thought of as setting up the same set of biochemical conditions as exist in at least the great majority of ectopic ossifications. The mechanical condition is therefore the ultimate but not the proximal cause of the ossification and these cases are thus brought into line with others in which no appeal can be made to a mechanical cause. There is also another interpretation which, although perhaps improbable, is not impossible. It may be that the tendon ossification should not be regarded as morphologically independent of the bone in which the tendon is inserted, the osteoblasts concerned having arisen in the tendon from indifferent cells, but that they are derived from osteoblasts of the bone which have migrated into the tendon. If the latter is the case, the tendon ossification is merely an extension of the bone, not a new differentiation, and is therefore rather a matter of architectural arrangement than of the origin of a new element. The same consideration may, of course, apply to ligaments.

Considering the facts which have been cited it may be

---

[1] Lordosis: a condition in which there is exaggeration of the normal curvature of the lumbar region of the spine.

concluded: (1) That special mechanical conditions are not essential either for the differentiation or for the continued activity of osteoblasts (*or* for the origin and continuation of osteogenesis). (2) Under certain conditions that remain obscure osteoblasts may be produced from (*or* osteogenesis may occur in) the connective tissue of any part of the body.

In other words, factors other than mechanical are responsible for the development of bone, although mechanical factors are paramount in determining the structural arrangement of the bone.

Section B. The Kinds of Mechanical Stressing influencing Bony Architecture

The development of the adaptive architecture of an adult bone proceeds in two stages. First is formed the architecture of the primary bone. This, as described in another section, is in replacing bones related rather to the pre-existing architecture of the cartilage than to mechanical stressing from without, but in membrane bones is probably determined by the arrangement of fibres and vessels in the connective tissue, and, perhaps, as Thoma and others hold for the bones of the skull, by direct response to the mechanical conditions. The development of the secondary or definitive architecture probably occurs directly or indirectly in response to the pattern of the mechanical stress.

The problem of the efficacy or otherwise of tension as a factor in the elaboration of bony architectures has already been discussed and, while the evidence is conflicting, it seems probable that greater importance must be attached to pressure. For this reason pressure only will here be considered, but it is likely that what is true of pressure is also true of tension, in so far as tension is effective at all.

A distinction must be drawn between constant and intermittent pressure. This was shown especially by the experiments of Jores (1920), the only investigator known to me who has advanced direct experimental evidence bearing on the subject of this section. Using young guinea-pigs and rabbits, Jores, by means of adhesive plaster, fixed bags containing water or mercury over the spine of the thoracic vertebrae behind the shoulder blades so that pressure was exerted through the skin on the spinous processes. After the lapse of a considerable time (up to more than 100 days) the animals were killed. The changes produced were seldom macroscopically visible and were confined to the ends of the spinous processes. It was nevertheless found that so long as the bone was subjected to constant unchanging pressure it underwent atrophy, but that removal of the pressure was followed by active growth of the bone, while alternating periods of pressure and of release from pressure gave especially clear increased growth of bone. The lengths of the alternating periods in the last case were twenty-four hours on and twenty-four hours off. An incidental result was to refute a contention of Roux that constant pressure does not cause atrophy when it is transmitted to the bone through cartilage, for in all or most of these experiments the ends of the spinous processes were still cartilaginous.

Loeschke and Weinnold (1922), in their study of the

skull to which reference has already been made, showed that resorption of bone from the inner side of the skull wall is associated with pressure exerted by the skull contents, while the deposition of bone is conditioned by the absence or reduction of the intra-cranial pressure. The weight of the brain increases up to twenty years, remains constant thereafter until forty or fifty, or may decrease slightly, decreases considerably from fifty to sixty and thereafter more rapidly (Boyd and Bischoff, from Loeschke and Weinnold). In a young growing skull both inner and outer walls are covered with secondary (periosteal) bone, but the *impressiones digitatae* regularly show resorption, and this may penetrate right through into the marrow cavity. Between twenty and thirty years the growth of both brain and skull comes to an end; the *impressiones digitatae* become rarer and shallower and interruption of the superficial lamellae by resorptive processes is uncommon. This is true of the roof of the skull, but on the sides and floor, where the brain, by virtue of its weight, still exerts a pressure, the *impressiones digitatae* are as well marked as before and still show resorption. In the third stage of the life-history, that of senile decrease of the volume of the brain, there is in the roof more active apposition of new bone on the inner face of the skull and the *impressiones digitatae* disappear, but on the floor of the skull resorption continues. But if, at any age, the intra-cranial pressure is increased, as by brain tumours, hydrocephaly, etc., resorption occurs on the inner surface and tumours of the brain may cause the formation of actual windows in the skull. The authors cite an interesting case of a boy who recovered from hydrocephaly but died of tuberculosis. The skull was at first expanded by the intra-cranial pressure and when this was relaxed there was a great thickening of the skull roof while its sides and floor still showed resorptive phenomena. The authors believe that pulsations of the brain make possible or assist the circulation of tissue fluids in the relatively rigid bone, but that where the brain exerts direct pressure on the skull wall, either by the force of its growth or by gravity, the pulsations are reduced or absent. It seems to the reader of their interesting work that this statement is not entirely clear: presumably it means that the pulsations, transmitted to the skull through the fluid bathing the brain, are damped by more direct pressure of the brain when this is in contact with the skull through the meninges. Constant, as distinct from intermittent, pressure, the authors believe increases the resistance to the flow of tissue fluids. Not all the capillaries in a tissue are at any one time carrying blood, but at different times the blood is carried through different capillary pathways. Increased resistance to the flow renders difficult this alternation between various capillary routes, the blood seeking a few larger channels, so that the tissue suffers in its nutrition and becomes atrophic. But the vascular system "undoubtedly" compensates for this disability and so modifies it as to achieve a "relative optimum" under the unfavourable conditions. If now the pressure be removed, the region concerned has a more efficient vascular apparatus than before it was applied, and will therefore grow more rapidly, until, by the regression of the new compensatory apparatus, equilibrium is restored. Thus Loeschke and Weinnold are in general agreement with Jores, differing from him only in regarding the two kinds of pressure as acting through the vascular system and not as a direct stimulus which becomes effective upon its removal.

One more point may be mentioned. Loeschke and Weinnold's vascular theory, which has just been outlined, clearly cannot apply to rapid alternations of pressure, such as are produced in the vertebral and appendicular skeletons by muscular exercise and in the skull by the arterial pulsations of the brain. Yet it is these pulsations that the authors have chiefly in mind as a factor furthering bone deposition on the inner side of the skull during the periods of relief from constant pressure by direct contact. There is no difficulty here, for two factors would act upon relief from the constant pressure: the presence of the compensating vascular apparatus built up during the period of pressure, and the previously suppressed arterial pulsations transmitted through the cerebro-spinal fluid, which are thought of as assisting the flow of tissue fluids.

Leriche and Policard (1928), discussing the work of Jores, interpret it in the sense of their humoral theory of ossification, considering that in the periods of pressure there is resorption of bone and the formation of a local calcific excess which is then, during the periods free from pressure, used for the formation of new bone. It is difficult to obtain a clear impression of the part these authors attribute to changes in the vascularity of ossifying tissues, but it is clear that they consider a defective circulation with stagnation of fluid in the tissue spaces to be closely associated with the colloidal infiltration which they regard as an essential phase of ossification. Greig (1931), who in principle accepts the views of Leriche and Policard, regards as two factors necessary for the formation of new bone an adequate blood supply and adequate calcium. Hyperaemia leads to de-calcification and rarefaction of bone, so that the supply of calcium locally available for new bone formation rises. Then, as the hyperaemia recedes and blood supply improves, the available calcium is built up again into bone, perhaps with a new architecture. Such hypotheses as these are not necessarily in conflict with that of Loeschke and Weinnold, for the resorption which these authors find during periods of steady pressure would provide the calcium used in the formation of new bone after the pressure has been removed, as Leriche and Policard suggest in interpreting the results obtained by Jores.

Weidenreich (1923) and Gebhardt, like Loeschke and Weinnold, regard the mechanical stimulus as mediated through the vascular system. These authors point out that Haversian systems always contain blood vessels, and Gebhardt (1905) remarks upon the existence of a relationship between the first bony trabeculae formed and the blood vessels of the earliest marrow. The trabeculae are always situated at a certain distance from the vessels, and Gebhardt believes that the vascular apparatus in a large degree determines the architecture of the bone. Weidenreich stresses the fact that lamellar bone always arranges itself around spaces, particularly spaces containing vessels, and that the form and size of the spaces is determined by their contents, i.e. by the axial vessels. When a vessel is narrowed, so, by the deposition of new bone, is the space in which it lies, and wide spaces contain many or wide vessels. Thus everything which changes the calibre or arrangement of the vessels becomes expressed also in the architecture of the bone: the calibre and arrangement of osteones depends on that of the vessels. But mechanical forces influence the calibre and arrangement of vessels very readily, tension causing them to become narrower and pulling them in the direction of the force. Thus those osteones which are laid down in this direction develop thicker walls and others remodel themselves so as to alter their direction. Pressure acts similarly; constant pressure, by obliterating the vessels, leads to atrophy of the bone, while intermittent pressure, causing alterations in the blood content of the vessels, is favourable.

An interesting and perhaps more probable suggestion than this is made by Greig. He points out that mechanical stressing, by producing miniature traumata in bone, would bring about locally a liberation of histamine. The histamine sets up a local hyperaemia which in its turn provokes a local decalcification and rarefaction. Thus the supply of locally available calcium is increased. As the hyperaemia recedes this is used in the formation of new bone, which is thus deposited in the region affected by the trauma, i.c. where it is needed.

Other authors, such as Petersen (1930) and Jores, think of the mechanical factor as directly stimulating the bone cells and osteoblasts and would regard the vascular architecture as dependent on that of the bone, and not *vice versa*. It is supposed that, where bone is suitably stressed, the osteoblasts are stimulated into activity, remaining quiescent elsewhere, while, under conditions leading to atrophy, osteoclasts bring about resorption of the bone.

It is of course impossible to decide between these theories; and there may be elements of truth in both. In favour of the second is an ingenious experiment by Glücksmann (unpublished). Having prepared tissue cultures in which osteogenesis was proceeding, he placed them, still *in vitro*, between two ribs taken out of embryos and still in part connected to one another by their inter-costal musculature. As time went on, the degenerating inter-costal musculature slowly contracted, so setting up pressure in the living osteogenic culture which lay between the ribs. Under these circumstances the orientation of the new bone formed depended on the extent to which bone already existing in the explant was calcified at the beginning of the experiment. If there was no, or very little, calcification in the explant the orientation was at right angles to the pressure, but if there was sufficient calcification to enable the explant to offer resistance to the pressure from the ribs, the long axes of the new trabeculae coincided with the direction of the pressure. This experiment at least shows a mechanical "architecture", on a very small scale it is true, developing in the absence of blood vessels. The mechanism is difficult to understand, and of course the experiment does not invalidate vascular theories of events *in vivo*, but suggests that it would be rash to reject altogether a possible direct response on the part of ossifying tissues to the stresses passing through them.

The possibility may here be mentioned that the decision as to whether bone is to be strengthened or resorbed may perhaps be made in the bone substance itself and not in the associated cells or in the surrounding medium; for it may be that bone, when unsuitably, insufficiently, or excessively stressed, undergoes some internal change, perhaps an intra-molecular rearrangement, which renders it liable to the depredations of osteoclasts. It is not without interest that osteoclasts have a considerable resemblance to the phagocytic multi-nuclear giant cells of pathology, whose function it is to destroy dead or degenerate cells or debris.

SECTION C

The work of Jores and of Loeschke and Weinnold leads us to conclude (with the reservation that more experimental evidence is strongly to be desired) that intermittent pressure leads to an increase in the total quantity of bone present, and constant pressure to a decrease, but, beyond the obvious suggestion that bone will be built up or destroyed according to the nature of the pressure exerted at any region, tells us nothing of the building up of an adaptive architecture. The embryonic bone is produced by the periosteum, which deposits bone as a veil covering the diaphysis, and by endochondral ossification, which produces a somewhat disorderly network of bone in the interior of the diaphysis. The form of the periosteal veil is determined directly by the form of the cartilaginous model which it covers, while the internal architecture seems to be determined by that of the pre-existing cartilage (Romeis, 1911) and by the arrangement of vessels in the primitive marrow. This vascular architecture is itself probably determined by the structure of the cartilage which it invades (Triepel, 1922a). Thus the endochondral bone, the first spongiosa, is trabecular and not compact because of its developmental relation to the pre-existing cartilaginous architecture. The early periosteal bone is from the first more compact than the endochondral bone, though of course less so than the adult compacta. The course of vessels in and below the periosteum, and perhaps the arrangement of its fibres, probably determines the trabecular architecture of the bone formed beneath it.

In later stages the spongy bone disappears from the middle of the diaphysis, which is thus a tube of compact bone, while in the epiphyses the state of affairs is reversed, the compact bone remains no more than a narrow rind and the whole central region or core becomes filled with a network of cancellous bone. The questions which now require to be answered are: why does the spongy bone disappear from the greater part of the diaphysis, and why does the compact bone remain thin in the epiphyses while the spongy bone is here most strongly developed? The answer to these questions has already been suggested in an earlier section; in the expanded ends of the bone the forces acting pass from the table-like ends through the inner parts of the epiphysial region to be concentrated in the periphery of the diaphysis. Thus in the diaphysis the inner part is free from stress, and it is well known that bone which is not subjected to stressing tends to atrophy, as in the bones of paralysed limbs or of limbs rendered functionless by amputation of essential parts, an atrophy doubtless to be correlated with the direct or indirect effect of the absence of the mechanical stimulus. In the epiphyses the core of bone is retained, because it is here subjected to the necessary stress, but this is not sufficient to explain its spongy architecture nor the failure of the region to develop a thick peripheral compacta. In the expanded epiphyses the stress is transmitted through a widened cross section, and therefore the stress per unit area is reduced. Now, the compacta has frequently been compared to a condensed spongiosa, the

spongiosa to a "feathered" compacta, and though Gebhardt (1910) showed that this comparison is not in all
respects acceptable, it is still true that both kinds of
structure consist of bone and spaces surrounded by bone.
The difference lies in the ratio of bone to inter-osseous
spaces. But since the stress is to be thought of as greater
per unit area in the compacta than in the spongiosa, it
follows that the stimulus to individual osteoblasts is likewise greater in the compacta, so that the amount of bone
deposited per osteoblast (or alternatively the number of
active osteoblasts per unit area) is greater in the compacta.
Hence the ratio of bone to inter-osseous spaces is greater
in the compacta. There are, however, still two other facts
to be considered. Firstly, the epiphysial bone originates as
a spongy core in the middle of the epiphysial cartilage and
is from the first net-like and not compact in structure
because of its origin in, and the primary dependency of its
architecture upon, the cartilage which it replaces. But
this does not explain the persistence of the cancellous
structure throughout life, for it has been shown by such
specimens as those illustrated in figs. 38 and 39 that, if the
mechanical conditions are changed, spongy bone can become transformed into compacta. Nevertheless, there is
doubtless a tendency for a structure which starts with a spongy
architecture to maintain it if the forces acting do not vary
beyond limits. Secondly, a fact which is frequently forgotten is that cancellous bone contains within its meshes a
marrow, and that this marrow is not just packing of no
importance but a definite organ of fat preservation or of
haematopoiesis. If the cancellous trabeculae went on
thickening until the condition approached that of compact

bone it is impossible to doubt that a pressure would be set up and that this would be a constant pressure such as that which, in the case of the brain pressing on the wall of the skull, leads to atrophy of the bone; thus the marrow would prevent. a wholesale transformation of cancellous into compact bone.

To explain the retention of the cancellous architecture in the epiphysial regions does not exhaust the problem, for that architecture is radically changed in form in the years after birth. Enough has already been said to make unnecessary any further extended discussion at this point. The architecture of the spongiosa in the epiphyses and the ends of the diaphyses is probably not at first mechanically adaptive but is determined by such factors as the pre-existing cartilaginous and vascular architectures. Subsequently, with the development of function (learning to walk, etc.), the architecture changes in the adaptive direction, the change being doubtless brought about by the direct or indirect effect of the changed stressing on the osteoblasts and osteoclasts.

It should perhaps be added in completion of this section that since the mechanical stressing of each part of the bone is determined partly by the general form of the bone, which is itself primarily settled by the self-differentiation of the cartilaginous model, so this primary self-differentiation is partly and indirectly responsible for the final structure of the bone.

Section D. The Mechanical Unit

The next step is to discover, if possible, a level in the organisation of bone at which the structure seen, while constituting the unit of which the mechanically adaptive gross architecture is built, cannot have been produced in response to the factors which call that architecture into being. The unit of structure found at this level corresponds to the bricks arranged, but not made, by the mechanical architect who builds the house.

Setting aside the non-lamellar bone of the early embryo and of certain special situations in the adult skeleton, bone has a lamellar structure, each lamella consisting of ground substance, in which salts are deposited, and regularly arranged bone fibrils. The lamellae are arranged concentrically around the whole bone (outer and inner general lamellae) or around blood vessels (Haversian systems or osteones). The osteones may be taken as the typical lamellar structure. They have been very thoroughly studied by Gebhardt (1905). There is no need to give a detailed repetition of his description nor an account of the reasoning which led him to his conclusions. It must suffice to say that he found the osteone to consist of concentrically arranged lamellae in each of which the fibrils were arranged in a spiral, that right- and left-handed spirals alternated in adjacent lamellae,[1] that he found variations in the steepness of the spiral windings and in the mode of combination in one osteone of lamellae whose fibrils were wound with different degrees of steepness; that he related

---

[1] Recent studies by Burkhardt (1929) show that the lamellar structure of osteones is actually less regular than Gebhardt supposed.

the different kinds of osteones produced by these variations to different mechanical functions, and that he regarded the arrangement and structure of the osteones as continuing on a finer scale the mechanical architecture of the coarser structures: the compacta and the trabeculae of the cancellous bone.

Here we meet with two problems: whether this last conclusion of Gebhardt's can be maintained, and whether the fine structure of the osteones, as distinct from their arrangement, can be produced by the same factors as determine the coarser architecture.

To begin with the first of the two questions, the adherents of the trajectorial theory have found themselves compelled to abandon Gebhardt's opinion, at least for the spongiosa. The reason for this is the necessity, for the survival of the theory, of showing that bone has a homogeneous structure. The trajectorial diagrams, to which the cancellous architecture is supposed to correspond, assume that the body under stress is homogeneous; if it were not homogeneous the diagrams would be different. Thus, if bone is not homogeneous, the trajectorial diagrams have only a dubious reference to it. The difficulty presented by the obviously heterogeneous structure of bone is evaded by the breccia theory of Petersen (1927), according to which the multitudinous anisotropies in the breccia cancel one another and so produce a statistical homogeneity. Thus the apparent value of osteone structure becomes meaningless, until it is rescued from this fate by a supposition of Benninghoff (1927). Accepting Petersen's breccia theory, Benninghoff suggests that the function of the osteone fragments is the taking up of stresses in such a manner as to maintain statistical homogeneity: a force impinging on the breccia meets with equal resistance from whatever direction it comes. Thus Benninghoff regards the functional significance of the osteone fragments, not as continuing on a finer scale the trajectorial architecture of the gross structure, but as providing a resistant mechanism which, by its disorderly arrangement, maintains the theoretically desired isotropy.

The breccia theory is probably true of the spongiosa, at any rate the histological structure is that of a conglomerate. But in the compact bone it can only be maintained with difficulty, if at all. Benninghoff's investigations of the lines of splitting reveal the existence of an orderly arrangement of parts. More direct evidence in the same sense is provided by measurements of the physical properties of compact bone. Rauber (figures from Kopsch) found that compact bone, subjected to tension in the direction of the long axis of the whole bone, was able to resist 9·25–12·41 kg. per mm.², subjected to pressure in the same direction resisted 12·56–16·8 kg. per mm.², while the corresponding figures for tension and pressure in a direction perpendicular to this were 4·8 kg. and 8·0 kg. per mm.². Thus compact bone is so far from having a homogeneous structure that its mechanical properties in two directions at right angles to one another are in the proportion of two to one. The significance of this depends on whether or not the arrangement of the osteones, which is the material basis of these differences in physical properties, is itself trajectorial or not. If the osteones are materialisations of the trajectorial lines in an ideal homogeneous body their arrangement can evidently not be used as an argument to support the objection based on the anisotropic structure of bone, for it is evident that that argument is only valid when the anisotropes do not harmonise with the disposition of trajectories. As stated in chapter IV Benninghoff (1925a, 1927) contends that the osteone arrangement in the compacta is trajectorial and smoothly continues the trajectorial structure of the cancellous bone, but Henckel (1931) disagrees; and until this matter has been definitely settled it will not be possible to decide on the significance of the anisotropic structure of compact bone.

These questions are difficulties only for the orthodox trajectorialist; from the standpoint of the present work they are not difficulties at all, for homogeneity of bone structure is not assumed. And this being so there is no need either to regard the structure of the osteones as mechanically meaningless or to rescue them from that oblivion by making them the servants of a homogeneity which does not exist at least in a large part of the bone. The first of the two questions is thus answered. Gebhardt (1905) having shown that osteones have a structure suited to resist mechanical stresses of certain kinds, and since Benninghoff has shown that the arrangement of osteones is, at least in a general way, such as would be mechanically expected, there is every reason for supposing that they do continue on a finer scale the mechanical architecture of the coarser structures.

Turning to the second question, it is clear that the spiral structure of osteones cannot be the embodiment of a trajectorial system, in the sense of embodying trajectories which preceded the development of the spiral structure. The trajectories in a body subjected to pressure or tension cannot be spiral unless the body already has a spiral structure into which the trajectories are led. The trajectories may indeed be spiral if a homogeneous body is subjected to a twisting force, but then the osteones should have a spiral structure in which all the fibrils would be wound one way, not reversing in alternate lamellae. If the direction of the twisting force were alternated, the fibrils should be arranged in two opposite directions in each lamella, and not in alternate lamellae; it is too much to believe that the periods of application of the torque in one direction should always coincide with the development of exactly one lamella. In Gebhardt's opinion (1905), osteone structure is indeed produced by mechanical factors but not by those responsible for the gross architecture of the bone. He points out that the fibrils in the lamellae are tangential to the axial vessel, and so are perpendicular to the radiating pressure pulses sent out by it. Fibrils are obviously tension-resistant and not pressure-resistant structures and the pressure pulses evidently subject them to tangentially directed tensions. The study of functional structures in the soft connective tissues has shown clearly that fibrils do arrange themselves along lines of tension and there can be little doubt that the same factor is concerned in the development of osteone structure. The change in fibrillar direction in alternate lamellae, Gebhardt suggests, may be determined by the elastic properties of a lamella with a single spiral winding. It is much less subject to deformation in the direction of its fibrils than in the direction at right angles to them. Thus a new lamella, formed upon the old as foundation, is during its development stressed in the second direction, and this tends to make the fibrils develop with this orientation, i.e. at right angles to those of the old lamella. That alternate directions of spiral winding are not at right angles to one another is due to the interference of other factors. Gebhardt points out that this idea is scarcely even a hypothesis, for the construction of which facts are lacking.

It must not be forgotten that such suggestions meet with a difficulty in the osteones with parallel fibres which are found in animals below the mammalian level (e.g. some of the bones of birds, like the fowl tibia). One would not expect such intimate mechanical details as those of which Gebhardt makes use in explaining the alternating spiral to differ greatly in homologous bones of different groups, and, since similar causes should produce similar effects, it is possible that some other factor is at work in producing this fundamental difference of structure.

Setting this aside, we have discovered, in the structural organisation at least of mammalian bones, a level at which the structure cannot be produced by the factors which call the gross architecture into being; that level is the structure of the osteone and of the lamellae of which it and other lamellar systems are composed.

This might suggest that the lamellar system or osteone would be the indivisible unit of bone structure, that every time such a structure undergoes modification it must be by the formation of whole new units and the destruction of whole old ones. But this is certainly not true, as is shown by the existence of the breccia, and is not a correct inference. The osteone is a unit only for the convenience of morphological description. That it is not more than this is shown by the absence of any precise and comprehensive definition; it is one of those terms of which everyone knows the meaning but which no one can define in such a manner that, in every specific case, there is no doubt about the exact limits of the structure to which reference is made. In a transverse section it is easy to point to an osteone and to indicate its boundaries, but in a section cut in the plane of the osteone the difficulty at once appears: osteones are branching and anastomosing structures, whose geography is that of the vessels they enclose, and no man can say (except quite arbitrarily) where one "unit" ends and another begins. Thus it is not the osteone which must represent the brick arranged by the builder-architect. Therefore, the unit must be at a lower level of structure. The individual lamella, like the osteone, and for the same reason, must be rejected. We are left with the concept "bone substance", consisting of fibrils, matrix, and calcareous impregnation. At this lowest of the histologically detectable structural levels, it is clear that none of the three components of bone substance can represent the unit, for neither fibre nor matrix nor calcium salt is by itself bone. The unit therefore must be these three together, that quantity of bone which cannot be further subdivided without ceasing to be bone. It must be this unit which, in obedience to those laws of which we know so little, is added here and subtracted there, producing at last a structure so well adapted to its task. Yet the admission of this introduces a new difficulty. We have three propositions: (*a*) that the gross architecture is mechanically adapted to the demands of function; (*b*) that it is made up of lamellar structures whose architecture is probably determined by mechanical forces of a kind different from those determining the gross architecture; and (*c*) that the lamellar structures, and therefore the gross architecture, are both built up of bone units. If the units are built up into the lamellae under the action of one set of factors, how does it come about that the sum of all the lamellae, the gross architecture, is adaptive to a different set of factors?

A number of answers can be made to this, and the answers do not exclude one another. Firstly, it can be said that natural selection would prevent the existence of mechanisms producing a fine structure which would add up to an inefficient gross structure. Secondly, in so far as the structure of bone is a breccia, the details of its fine structure are unimportant so long as the bone units are put together in a mechanically stable manner. Thirdly, the factors especially thought to influence the fine structure, principally the blood vessels, are themselves of a kind likely to be affected by the mechanical stressing of a bone as a whole, so that a measure of coincidence between the effects of the two sets of factors is to be expected. The stresses passing through the bone determine where there is to be more bone, where less, but the detailed structure of the lamellae formed is determined by the intimate mechanical conditions set up locally by such factors as pressure pulses from blood vessels.

CHAPTER VII

SUMMARY AND CONCLUSIONS

Although in the preceding pages it has not been found possible to adhere to the strict form of the trajectorial theory, few will doubt that bone structure is at least in a very large degree determined, directly or indirectly, by reaction to the mechanical forces which it is the function of bone to resist. Nevertheless, it will have become clear from the facts presented that the time is past when it was possible to describe the whole long and complex story of skeletogenesis in terms of the effects of mechanical stresses, whether of extrinsic and functional origin or arising in direct consequence of the developmental process itself. The form of the cartilaginous model is produced, at least in the case of the larger shafted elements, under the aegis of a growth pattern intrinsic from a very early stage within each element. Neither the functioning of muscles, nor mechanical interaction between the developing elements, can now be allowed any major role in the first production of form, even though the second set of factors does play an essential part in bringing about the separation of the continuous anlage into the several elements of the limbs. But here must be introduced one of the many qualifications which are rendered necessary by the very incomplete state of knowledge: practically nothing is known, in these respects, of cartilaginous elements outside of the limbs of one embryo, the chick, and practically nothing of the development of the form of the vertebrae or of the skull, and little of the production of such concave articular structures as the acetabulum. In cartilaginous elements of another kind, such as the otic, optic and nasal capsules, which enclose and protect softer parts, the form is apparently determined wholly by the organ encapsuled. This difference between the two types of cartilaginous elements is of course to be expected, for while the form of a limb must evidently be determined by the system of levers which are its essential parts and the reason for its existence, in the case of the sense organs cartilage is clearly secondary, and the sense organ itself the partner which must be allowed to determine the form.

Even in the primary stage of development, factors of a mechanical kind are not quite without importance, and instances have been given in which it appears that the full perfection even of the cartilage model cannot be achieved by the action of the intrinsic pattern of growth alone. But it is in the bony skeleton that the influence of intrinsic form-determining factors sinks lowest, and perhaps to nothing. The bony skeleton of the adult shows, both in its normal structure and after modification, so close a relation to mechanical needs and so great a power of apparently adaptive change, that its mechanically suitable nature cannot be doubted. The harmony with functional demands is traceable at all levels of its structure: in the shape of the whole bone, in its division into compact and cancellous bone, in the arrangement of the osteones of the first and of the trabeculae of the second, in the structure of individual osteones and of other lamellar systems, and finally in the intimate composition of the substance bone itself. But although at all levels there can be shown a relationship between structure and function which is well expressed by such terms as "suitability" or "harmony", it cannot be assumed that the relationship is always one of cause and effect, for many different causes combine to produce the finished architecture, and not, as is often stated or implied, only the single one of functional stress.

The position taken in this work is not that of the orthodox trajectorial theory. That theory regards the trabeculae of bone as embodying trajectories in a body of the same form as the bone, stressed in the same way, and of homogeneous composition. The trabeculae are thought of as drawn in such an ideal and hypothetical foundation, and the theory is intimately bound up with the existence of orthogonal structures in which all the trabeculae must cross one another at right angles, a state of affairs which, as stated in an earlier section, is not always found in nature. The view here expressed assumes no such relationship to an ideal homogeneous body, is not affected by the decidedly anisotropic structure of bone, and is compatible with the absence of orthogonal structure. It regards the form and structure of bone as a compromise between many factors. One of these is the original dependency of its form, if it is a replacing bone, on that of the self-differentiating cartilage model; if it is a membrane bone, on the mechanical conditions under which it underwent its early development, conditions which may have ceased to exist with the ending of the embryonic period. Another is the influence of stresses due to functional activity in changing both its form and its structure, as in the post-foetal development of the human calcaneus. A third is the need for compromise with neighbouring organs, as the contours of the human tibia are moulded by the lateral pressure of muscles lying against it. A fourth is the influence upon later structure of the earlier forms which it has superseded, such as the relation between the early cancellous bone and the architecture of the cartilage which it has displaced, a relation which is perhaps still expressed, even in the adult, in the fine meshed bone of the epiphyses. A fifth is probably the necessity for the organism of respecting the needs of the marrow. A sixth is the influence of more or less passive changes of form on the results of modes and patterns of growth which have not been changed, as in bent elements in chorio-allantoic grafts and chondro-dystrophic embryos. At a lower level of organisation, the structure of the osteone and of other lamellar systems, reasons have been given for considering that while it is at least to some extent the product of mechanical causes, it is not to the stresses produced by function, but rather to intimate mechanical relationships within the tissue, that the characteristic structures owe their origin, to such factors as the intermittent waves of pressure sent out by pulsating blood vessels and the like. Among all these factors it is clear that one, the direct or indirect influence of modes of mechanical stressing, is predominant in the secondary phase of development and in later re-constructions, but the multiplicities of factors modifying the influence of this should prevent us from expecting a precise theoretically perfect adaptation to the demands of this single determinant. When a bony structure is recon-structed following a change in the pattern of stress within it, the new architecture develops, not in pure, direct, and unalloyed dependency on that pattern, but in dependency also on local histological and histo-chemical conditions, as on the vascular anatomy and on the distribution of biochemical conditions favourable to ossification. It is probable that these determinants are themselves more or less influenced by the pattern of stress, as Weidenreich suggests for the blood vessels, but it is hard to believe that their geography can always be such that their influence on the developing bone is in perfect accord with the demands made by function upon it. The new vascular architecture is developed by modification of the old, and the old, which is not in adaptation to the changed conditions, must therefore influence the new.

Every bony structure is a compromise and no compromise is perfect. The trajectorial theory in its original and most rigid form played a useful part in calling attention to the relation which obviously does exist between bony structures and patterns of stress, but like many theories it went too far when it claimed all power for its own particular factor and neglected all others. It requires, not rejection, but dilution.

So far the ground is fairly firm, but the attempt to pass beyond this point is to venture into a quagmire little better than speculation with very few established facts to act as signposts.

When a bony architecture is produced, or modified, in causal relationship with functional stresses, it may suffer change at all levels of its organisation. Modification of a bone involves the addition and subtraction of material, and discussion of the unit added and subtracted showed that this was not the morphologists' unit, the osteone, but was at a lower level than the osteone, the unit of bone substance itself.

It is the question of the kind of functional stimulus which is effective and the manner of its action that remain as the most obscure problem in skeletogenesis. The majority of authors, following the lead of the founders of the trajectorial theory, attribute equal efficacy to both pressure and tension stresses; others, bringing forward strong evidence for rejecting tension and relying entirely on pressure, are met by Weidenreich's work on tendon ossification. Perhaps Triepel is right when he says one should not distinguish between pressure and tension, because each is always and inevitably accompanied by the other, but should think simply of a state of stress. Of the mode of stressing, as distinguished from the kind of stress, a very little more can be said; the work of Jores and of Loeschke and Weinnold suggests that intermittent pressure is favourable, and constant pressure unfavourable, to the deposition of bone. Even if this is true, however, it is not entirely clear whether an influence on the rate of growth or resorption of bone can be fairly regarded as also an influence on the kind of architecture developed.

Whether mechanical stresses act directly on the osteogenic process, or indirectly by first influencing some other organ or process, is as obscure as the problem of the kind of effective stress and the mode of stressing. Petersen sees in the network of bone corpuscles a system of cells, extending throughout the bone, which can respond directly to the tiny deformations caused by mechanical forces and can set up in suitable places the processes of deposition and resorption of bone, but other authors, like Weidenreich, envisage an indirect action which, by influencing the calibre and distribution of blood vessels, alters the conditions of nutrition or of any other factor depending on the vessels and by this means brings about changes in bone structure. Triepel, while recognising that mechanical forces do influence bone structure, puts forward his new principle of "harmonische Einfügung", contending that the main factor in determining internal structure is external form; and striking evidence in favour of an idea related to this, but depending not on form and an abstract principle, but on form and a definitely known mode of normal development, was found in the bent elements of chondro-dystrophic embryos and of normal chorio-allantoic grafts.

Like every other discussion of a living problem in science, this work ends, metaphorically, not with a full stop but with a question mark; for it is to be hoped that, whatever its faults, it helps to show in what direction further exploration may best succeed in colouring the blank spaces on the map. While further studies of bony architectures and their correlations with functional requirements will be interesting, the real need is not for these, but for investigations of the mechanism which underlies the adaptations known to occur. Whether the production of functionally suitable structures by the effect of passive changes of form on the results of normal processes of growth and development should be regarded as one of the major mechanisms involved in "adaptive" changes or as merely an interesting but isolated special case, what the relation is between bony structures and vascular patterns, whether mechanical forces act directly on the osteogenic process or indirectly through some intermediate mechanism, are all problems for the future, but they are problems whose answers would throw a flood of light on the obscure processes of skeletogenesis and which are probably not inaccessible to the ingenious mind and skilful technique of the biological experimenter.

REFERENCES

* *These are works which I have not seen.*

† *These are works containing useful general discussions, chiefly on the mechanical structure of bone.*

APPLETON, A. B. (1922). "Effect of abnormal posture upon the muscles and limbs of a healthy animal." *Proc. Anat. Soc.* p. 41.

—— (1925 a). "Influence of mechanical factors on epiphysial ossification." *Proc. Anat. Soc.* p. 30.

—— (1925 b). "Experiments concerning the morphogenesis of bone." *Proc. Anat. Soc.* p. 81.

BENNINGHOFF, A. (1924). "Experimentelle Untersuchungen über den Einfluss verschiedenartiger mechanischer Beanspruchung auf den Knorpel." *Verh. Anat. Ges.* Vers. 33, p. 194.

—— (1925 a). "Spaltlinien am Knochen, eine Methode zur Ermittlung der Architektur platter Knochen." *Verh. Anat. Ges.* Vers. 34, p. 189.

—— (1925 b). "Der funktionelle Bau des Hyalinknorpels." *Ergeb. Anat. Ent.-gesch.* **26**.

—— (1927). "Über die Anpassung der Knochencompakta an geänderte Beanspruchung." *Anat. Anz.* **63**, 289.

—— (1930). "Über Leitsysteme der Knochencompakta. Studien zur Architektur der Knochen, 3." *Morphol. Jahrb.* **65**, 11.

—— (1934). "Die Architektur der Kiefer und ihre Weichteilbedeckung." *Paradentium,* Jahrg. 6, p. 1.

†—— (1935). "Form und Funktion 1." *Zeitschr. ges. Naturwiss.* p. 149.

BERNAYS, A. (1878). "Die Entwicklungsgeschichte des Kniegelenkes des Menschen mit Bemerkungen über die Gelenke im Allgemeinen." *Morphol. Jahrb.* **4**.

BERNHARD, F. (1924). "Über den Einfluss der Muskulatur auf die Formgestaltung des Skelettes." *Roux Arch.* **102**, 489.

†BIEDERMANN, W. (1913). "Physiologie der Stütz- und Skelettsubstanzen," in Winterstein's *Handbuch der vergleichenden Physiologie,* **3**, 1, 1.

BIER, A. (1923). "Über Knochenregeneration, über Pseudarthrosen und über Knochentransplantate." *Zeitschr. klin. Chir.* **127**.

BOND, C. J. (1913–14). "On the late result of three cases of transplantation of the fibula: with remarks on the process of growth and the physiological development of transplanted bone." *Brit. J. Surg.* **1**, 610.

BRANDT, W. (1927). "Schultergürteluntersuchungen an transplantierten Gliedmassen bei *Triton taeniatus.*" *Roux Arch.* **112**, 149.

BRASH, J. C. (1934). "Some problems in the growth and developmental mechanics of bone." *Edin. Med. J.* New Series, **41**, 305.

BRAUS, H. (1909). "Gliedmassenpropfung und Grundfragen der Skelettbildung. 1. Die Skelettanlage vor Auftreten des Vorknorpels und ihre Beziehung zu den späteren Differenzierungen." *Morphol. Jahrb.* **39**, 155.

—— (1910). "Angeborene Gelenkveränderungen bedingt durch künstliche Beeinflussung des Anlagematerials." *Roux Arch.* **30**, 459.

BROILI (1922). "Über den feineren Bau der verknöcherten Sehnen (verknöcherten Muskeln) von *Trachodon.*" *Anat. Anz.* **55**, 465.

BRUNST, V. (1927). "Zur Frage nach dem Einfluss des Nervensystems auf die Regeneration." *Roux Arch.* **109**, 41.

—— (1932). "Über die Bedeutung der Funktion für die definitive Skelettbildung der Extremitäten bei der Regeneration." *Roux Arch.* **125**, 640.

BURKHARDT (1929). "Über den Aufbau der menschlichen Osteone." *Anat. Anz.* **67**, 96.

BURR, H. S. (1930). "Hyperplasia in the brain of *Amblystoma.*" *J. Exp. Zool.* **55**.

BUSSE and BLECHER (1904). "Über myositis ossificans." *Dtsch. Zeitschr. Chir.* **73**, 388.

CAREY, E. J. (1921). "Studies in the dynamics of histogenesis. Compression between accelerated growth centres of the segmental skeleton as a stimulus to joint formation." *Amer. J. Anat.* **29**, 93.

—— (1922). "Direct observations on the transformation of the mesenchyme in the thigh of the pig embryo (*Sus scrofa*), etc." *J. Morphol.* **37**.

—— (1929). "Studies in the dynamics of histogenesis. Experimental, surgical, and roentgenographic studies of the architecture of human cancellous bone, the resultant of back pressure vectors of muscle action." *Radiology*, **13**, 3.

CAREY, E. J., ZEIT, and McGRATH (1927). "The regeneration of the patellae of dogs." *Amer. J. Anat.* **40**, 127.

*Christen. "Die Entstehung der Diaphysenbrüche." *Verh. dtsch. Orthop. Ges.* Kongress 13, p. 12.

Colton, H. S. (1929). "How bipedal habit affects the bones of the hind legs of the albino rat." *J. Exp. Zool.* **53**.

Dixon, F. (1910). "The architecture of the cancellous tissue forming the upper end of the femur." *J. Anat. and Physiol.* **44**, 223.

Dowgjallo, N. D. (1932). "Die Struktur der Compacta des Unterkiefers bei normalen und reduzierten Alveolarfortsatz." *Zeitschr. Anat. Ent.-gesch.* **97**, 55.

Fell, H. B. (1925). "The histogenesis of cartilage and bone in the long bones of the embryonic fowl." *J. Morphol. and Physiol.* **40**, 417.

—— (1928). "Experiments on the differentiation *in vitro* of cartilage and bone, Part I." *Arch. exp. Zellforsch.* **7**, 390.

—— (1932). "The osteogenic capacity *in vitro* of periosteum and endosteum isolated from the limb skeleton of fowl embryos and young chicks." *J. Anat.* **66**, p. 157.

—— (1933). "Chondrogenesis in cultures of endosteum." *Proc. Roy. Soc.* B, **112**, 417.

Fell, H. B. and Canti, R. G. (1934). "Experiments on the development *in vitro* of the avian knee joint." *Proc. Roy. Soc.* B, **116**, 316.

Fell, H. B. and Robison, R. (1929). "The growth, development, and phosphatase activity of embryonic avian femora and limb-buds cultivated *in vitro*." *Biochem. J.* **23**, 767.

*Fick, L. (1857). *Über die Ursachen der Knochenform.* Göttingen.

Fick, L. and Fick, A. (1859). "Über die Gestaltung der Gelenkflächen." *Arch. Anat. Physiol.*

Fick, R. (1890). Über die Form der Gelenkflächen." *Arch. Anat. Physiol.* Anat. Abt. 1890, Suppl. Bd. p. 391.

—— (1921). "Über die Entstehung der Gelenkformen." *Abh. Preuss. Akad. Wiss.* Jahrg. 1921, Phys.-Math. Kl.

—— (1928). "Zur Erinnerung an H. v. Meyer und Bemerkungen über Knochen und Gelenkformung (Eröffnungsansprache)." *Verh. Anat. Ges.* Vers. 37, p. 3.

Fuld, E. (1901). "Über Veränderungen der Hinterbeinknochen von Hunden infolge Mangels der Vorderbeine." *Roux Arch.* **11**.

Gask, G. E. (1913–14). "Autoplastic graft of fibula into humerus after resection for chondro-sarcoma, with observation on bone grafting." *Brit. J. Surg.* **1**, 49.

Gebhardt, F. (1905). "Über funktionelle wichtige Anordnungsweise der groberen und feineren Bauelemente des Wirbeltierknochens. 2. Spezieller Teil." *Roux Arch.* **20**, 187.

GEBHARDT, F. (1910). "Die spezielle funktionelle Anpassung der Röhrenknochen-Diaphyse." *Roux Arch.* **30**, 516.

GREIG, D. M. (1931). *Clinical Observations on the Surgical Pathology of Bone.*

*v. GUDDEN (1874). *Experimentaluntersuchungen über das Schädelwachstum.* München.

HAMBURGER, V. (1928). "Die Entwicklung experimentell erzeugter nervenloser und schwach innervierter Extremitäten von Anuren." *Roux Arch.* **114**, 272.

HEIDSIECK (1934). "Eine Modellvorstellung vom Knorpel." *Anat. Anz.* **78**, 175.

HENCKEL, K. O. (1931). "Vergleichend-anatomische Untersuchungen über die Struktur der Knochencompacta nach der Spaltlinienmethode." *Morphol. Jahrb.* **66**, 22

HENKE, W. and REYHER, C. (1874). "Studien über die Entwicklung der Extremitäten des Menschen, insbesondere der Gelenkflächen." *Sitz. Ber. kais. Akad. Wiss.* Math.-naturwiss. Kl. 70, Jahrg. 1874. Wien, 1875.

HESSER, C. (1925). "Beiträge zur Kenntnis der Gelenkentwicklung beim Menschen." *Morphol. Jahrb.* **55**, 489.

*HIRSCH, H. (1895). *Die mechanische Bedeutung der Schienbeinform.* Berlin.

HUGGINS, C. B. (1931). "The formation of bone under the influence of epithelium of the urinary tract." *Arch. Surg.* **22**, 377.

HUGGINS, C. B. and SAMMETT, J. F. (1933). "Function of the gall bladder epithelium as an osteogenic stimulus and the physiological differentiation of connective tissue." *J. Exp. Med.* **58**, 393.

JACOBSON, W. (1932). "Über die Zellvorgänge in der ersten Entwicklungsstadien des knorpel- und knochenbildenden Gewebes." *Verh. Anat. Ges.* Vers. 41, Lund, p. 186.

JANSEN, MURK (1920). *On Bone Formation.*

JENNY, H. (1912). "Notizen über ein männliches Schaf ohne vordere Extremität." *Anat. Anz.* **40**.

JORES, L. (1920). "Experimentelle Untersuchungen über die Einwirkung mechanischen Druckes auf den Knochen." *Zieglers Beitr. path. Anat. allg. Path.* **66**, 433.

KOCH, J. C. (1917). "Laws of Bone Architecture." *Amer. J. Anat.* **21**, 177.

KOCH, W. (1926–27). "Über verknöcherte Sehnen bei Macropus." *Anat. Anz.* **62**, 138.

KOKOTT, W. (1933). "Über den Bauplan des fötalen Hirnschädels." *Beitr. Anat. functioneller Systeme,* **I**, 471.

KOPSCH, FR. (1922). Rauber's *Lehrbuch und Atlas der Anatomie des Menschen*, Abt. 2: Knochen, Bänder.

LANDAUER, W. (1927). Untersuchungen über Chondrodystrophie. I. Allgemeine Erscheinungen und Skelett chondrodystrophischer Hühnerembryonen." *Roux Arch.* **110**, 197.

——— (1929). "Funktionelle Strukturen von Knorpel und Knochen und ihre Entstehung." *Roux Arch.* **115**, 911.

LEBEDINSKY, N. G. (1925a). "Entwicklungsmechanische Untersuchungen an Amphibien. 2. Die Umformung der Grenzwirbel bei *Triton taeniatus* und die Isopotenz allgemein homologer Körperteile des Metazoenorganismus." *Biol. Zentralbl.* **45**.

——— (1925b). "Die Isopotenz allgemein homologer Körperteile des Metazoenorganismus." *Abh. theoret. Biol. von Julius Schaxel*, Heft 22.

LERICHE, R. and POLICARD, A. (1928). *The normal and pathological physiology of bone*. Trans. by S. Moore and J. A. Key. London.

LESSHAFT (1892). "Ueber das Verhältnis der Muskeln zur Form der Knochen und der Gelenke." *Verh. Anat. Ges.* Vers. 6, Wien, p. 178.

*LEVI, G. M. (1930). "Neoformazione sperimentale di cartilagine secondaria nella cavia e nel pulcino." *Boll. Soc. Biol. sper.* **5**, 117.

LEXER, E. (1922). "Über die Entstehung von Pseudarthrosen nach Frakturen und nach Knochentransplantationen." *Arch. klin. Chir.* **119**, 520.

LOESCHKE and WEINNOLD (1922). "Über den Einfluss von Druck und Entspannung auf das Knochenwachstum des Schädels." *Zieglers Beitr. path. Anat. allg. Path.* **70**, 406.

LUBOSCH, W. (1910). *Bau und Entstehung der Wirbeltiergelenke*. Gustav Fischer, Jena.

MAIER, O. (1930). "Zur Frage der Entstehung der Gelenkform (an Hand eines Falles von pseudarthrotischen Kugelgelenkes)." *Anat. Anz.* **70**, 278.

MAIR, R. (1926). "Untersuchungen über die Struktur der Schädelknochen." *Zeitschr. mikr.-anat. Forsch.* **5**, 625.

MARTIN, B. (1933–34). "Zur Entstehung der sympathischen Knochenerkrankung." *Arch. klin. Chir.* **178**, 81.

MAXIMOW (1907). "Experimentelle Untersuchungen zur postfötalen Histogenese des myeloiden Gewebes." *Zieglers Beitr. path. Anat. allg. Path.* **41**, 122.

MEYER, H. v. (1867). "Die Architektur der Spongiosa." *Arch. Anat. Physiol.*

MOLLIER (1910). "Ueber Knochenentwicklung." *Sitz. ber. Ges. morphol. Physiol. in München.*

MÜLLER, W. (1921). "Experimentelle Untersuchungen über extra-artikuläre Knochenüberbrückung von Gelenken." *Bruns. Beitr. klin. Chir.* **124**, 315.

—— (1922). Same title. *Zentralbl. Chir.* **49**, 553 (summary of previous paper).

MURRAY, P. D. F. and HUXLEY, J. S. (1925). "Self-differentiation in the grafted limb-bud of the chick." *J. Anat.* **59**, 379.

MURRAY, P. D. F. (1926). "An experimental study of the development of the limbs of the chick." *Proc. Linn. Soc. N.S.W.* **51**, 187.

—— (1928). "Chorio-allantoic grafts of fragments of the two-day chick, with special reference to the development of the limbs, intestine and skin." *Australian J. Exp. Biol. Med. Sci.* **5**, 237.

MURRAY, P. D. F. and SELBY, DORIS (1930). "Intrinsic and extrinsic factors in the primary development of the skeleton." *Roux Arch.* **122**, 629.

NATANSON, K. (1903). "Knorpel in der Niere." *Wien. klin. Wochenschr.* p. 857.

NAUCK, E. TH. (1926). "Beiträge zur Kenntnis des Skelettes der paarigen Gliedmassen der Wirbeltiere. 1. Die Ontogenese der Gelenkflächen und die Morphologie des Vordergliedmassen-gürtels der Wirbeltiere." *Morphol. Jahrb.* **56**, 22.

NIVEN, J. S. F. (1933). "The development *in vivo* and *in vitro* of the avian patella." *Roux Arch.* **128**, 480.

OGAWA, C. (1926). "Einige Experimente zur Entwicklungsmechanik des Amphibienhörbläschen." *Fol. Anat. Jap.* **4**.

PETERSEN, H. (1927). "Über den Feinbau der menschlichen Skelett-teile." *Roux Arch.* **112**, 112.

†—— (1930). "Die Organe des Skelettsystems" in v. Möllendorff's *Handbuch der mikroskopischen Anatomie des Menschen*, **2**, Part 2.

POSCHARISSKY, J. F. (1905). "Über heteroplastische Knochenbildung." *Zieglers Beitr. path. Anat. allg. Path.* **38**, 135.

ROGGEMANN, H. (1930). "Untersuchungen über die Heilung von Knochenbrüchen bei Vögeln." *Zeitschr. wiss. Zool.* **137**, 627.

ROMEIS, B. (1911). "Die Architektur des Knorpels vor der Osteogenese und in der ersten Zeit derselben." *Roux Arch.* **31**.

†ROUX, W. (1885). "Beiträge zur Morphologie der funktionellen Anpassung. 3. Beschreibung und Erläuterung einer knöchernen Kniegelenksankylose." *Arch. Anat. Physiol.* Anat. Abt. **1885**, and *Roux Ges. Abhandl.* **1895**.

188   REFERENCES

†Roux, W. (1893). "Kritisches Referat über 'Das Gesetz der Trans-
formation der Knochen' von Julius Wolff." *Berliner klin. Wochenschr.*
and *Roux Ges. Abhandl.* 1895.
SACERDOTTI, C. and FRATTIN, G. (1902). "Über die heteroplastische
Knochenbildung." *Virchows Arch. path. Anat. Physiol.* **168**, 431.
SCHAFFER, J. (1911). "Trajektorielle Strukturen im Knorpel." *Verh.
Anat. Ges.* **25**, 162.
SCHMALHAUSEN, J. (1925). "Über die Beeinflussung der Morphogenese
der Extremitäten vom Axolotl durch verschiedene Faktoren."
Roux Arch. **105**, 483.
SCHMIDT, R. (1899). "Vergleichend-anatomische Studien über den
mechanischen Bau der Knochen und seine Vererbung." *Zeitschr.
wiss. Zool.* **65**, 65.
SCHULIN, K. (1879). "Über die Entwicklung und weitere Ausbildung
der Gelenke des menschlichen Körpers." *Arch. Anat. Physiol.*
Anat. Abt.
SEDILLOT (1864). "De l'influence des fonctions sur la structure et la
forme des organes." *C.R. Acad. Sci.* **59**, 539.
STIEVE, H. (1927). "Versuche über die Tätigkeitsanpassung langer
Röhrenknochen. 1. Der Einfluss stärkerer Inanspruchnahme
auf die Länge und Dicke der Mittelfussknochen und Zehenglieder
am Hinterlaufe des Kaninchens." *Roux Arch.* **110**, 528.
STUDITSKY, A. N. (1934a). "Experimentelle Untersuchungen über
die Histiogenese des Knochengewebes. 2. Über die Bedeutung
der Wechselwirkung des Knorpelgewebes und des Periostes nach
den Ergebnissen der Kulturen in der Allantois." *Zeitschr.
Zellforsch. mikr. Anat.* **20**, 636.
—— (1934b). "The mechanism of the formation of regulating
structures in the embryonic skeleton." *C.R. Acad. Sci. U.R.S.S.*
p. 637.
—— (1935). "Experimentelle Untersuchungen der Histogenese des
Knochengewebes. 6. Ueber den Mechanismus der formbildenden
Prozesse im Embryonalskelett." *Arch. Russes d'Anat., d'Histol.
et d'Embryol.* p. 311. (Russian; German translation of discussion
and summary.)
THOMA, R. (1911). "Untersuchungen über das Schädelwachstum und
seine Störungen. 1. Die Spannung der Schädelwand." *Virchows
Arch. path. Anat. Physiol.* **206**, 201.
—— (1913). "Untersuchungen über das Schädelwachstum und
seine Störungen. 2. Das fetale Wachstum." *Virchows Arch. path.
Anat. Physiol.* **212**, 1.

†Thompson, D'Arcy, W. (1917). *Growth and Form*, Chapter XVI.

*Triepel, H. (1902). *Einführung in die physikalische Anatomie*. Wiesbaden.

—— (1904). "Architekturen der Spongiosa bei abnormer Beanspruchung der Knochen." *Anat. Hefte*, **25**, 209.

—— (1922 a). "Die Architektur der Knochenspongiosa in neuer Auffassung." *Zeitschr. Konstitutionslehre*, **8**, 269.

—— (1922 b). "Über gestaltliche Beziehungen zwischen Struktur und Organform." *Zeitschr. Anat. Ent.-gesch.* **63**, 608.

—— (1922 c). *Die Architekturen der menschlichen Knochenspongiosa*. München and Wiesbaden.

Troitsky, W. (1932). "Zur Frage der Formbildung des Schädeldaches." *Zeitschr. Morphol. Anthrop.* **30**, 504.

Twitty, V. C. (1932). "Influence of the eye on the growth of its associated structures, studied by means of heteroplastic transplantations." *J. Exp. Zool.* **61**.

Voss, H. E. V. (1926). "Weitere Beobachtungen über metaplastische Knochenbildung im Ovarialtransplantat." *Biol. Generalis*, **2**, 1.

Ward (1838). *Human Osteology*. London.

Weidenreich, F. (1922). "Über formbestimmende Ursachen am Skelett und die Erblichkeit der Knochenform." *Roux Arch.* **51**, 436.

—— (1923). "Knochenstudien. 2. Über Sehnenverknöcherungen und Faktoren der Knochenbildung." *Zeitschr. Anat. Ent.-gesch.* **69**, 558.

—— (1924). "Wie kommen funktionelle Anpassungen der Aussenform des Knochenskelettes zustande?" *Paläontol. Zeitschr.* **7**, 34.

Weinnold, H. (1922). "Untersuchungen über das Wachstum des Schädels unter physiologischen und pathologischen Verhältnissen." 1 and 2. *Zieglers Beitr. path. Anat. allg. Path.* **70**, 311, 345.

Wermel, J. (1934). "Untersuchungen über die Kinetogenese und ihre Bedeutung in der onto- und phylogenetischen Entwicklung (Experimente und Vergleichungen an Wirbeltierextremitäten). 1. Allgemeine Einleitung, Veränderungen der Länge der Knochen." *Morphol. Jahrb.* **74**, 143.

—— (1935 a). "Untersuchungen etc. 2. Veränderungen der Dicke und der Masse der Knochen." *Morphol. Jahrb.* **75**, 92.

—— (1935 b). "Untersuchungen etc. 3. Veränderungen der Widerstandsfähigkeit der Knochen." *Morphol. Jahrb.* **75**, 128.

—— (1935 c). "Untersuchungen etc. 4. Lageveränderungen der Skelettelemente und damit verbundene Veränderungen der Knochen- und Gelenkform." *Morphol. Jahrb.* **75**, 180.

190   REFERENCES

WERMEL, J. (1935*d*). "Regeneration der Knochen und Gelenke, sowie Neubildung der letzteren." *Morphol. Jahrb.* **75**, 445.

—— (1935*e*). "Untersuchungen etc. 6. Veränderungen der Muskulatur." *Morphol. Jahrb.* **75**, 452.

—— (1935*f*). "Untersuchungen etc. 7. Veränderungen der Haut und der Hautgebilde, auch Schluss der Untersuchungen." *Morphol. Jahrb.* **75**, 459.

WILLICH, C. TH. (1924). "Experimentelles über Knochenregeneration und Pseudarthrosenbildung." *Arch. klin. Chir.* **129**, 203.

WOLFF, J. (1870). "Ueber die innere Architektur der Knochen und ihre Bedeutung für die Frage vom Knochenwachstum." *Virchows Arch. path. Anat. Physiol.* **50**, 389.

—— (1892). *Das Gesetz der Transformation der Knochen.* Berlin.

—— (1899). "Die Lehre von der funktionellen Knochengestalt." *Virchows Arch. path. Anat. Physiol.* **155**, 256.